Mathematical Biology

Pergamon Titles of Related Interest

KARNOPP Bond Graph Techniques for Dynamic Systems in Engineering and Biology

KARREMAN Cooperative Phenomena in Biology

Related Journals*

ANNALS OF BIOMEDICAL ENGINEERING
BIORHEOLOGY
BULLETIN OF MATHEMATICAL BIOLOGY
CELLULAR AND MOLECULAR BIOLOGY
INTERNATIONAL ABSTRACTS OF BIOLOGICAL SCIENCES
MATHEMATICAL MODELLING

***Free specimen copies available upon request.**

Mathematical Biology

A Conference on Theoretical Aspects of Molecular Science

Edited by T.A. Burton

Pergamon Press

NEW YORK • OXFORD • TORONTO • SYDNEY • PARIS • FRANKFURT

Pergamon Press Offices:

U.S.A.	Pergamon Press Inc., Maxwell House, Fairview Park, Elmsford, New York 10523, U.S.A.
U.K.	Pergamon Press Ltd., Headington Hill Hall, Oxford OX3 0BW, England
CANADA	Pergamon Press Canada Ltd., Suite 104, 150 Consumers Road, Willowdale, Ontario M2J 1P9, Canada
AUSTRALIA	Pergamon Press (Aust.) Pty. Ltd., P.O. Box 544, Potts Point, NSW 2011, Australia
FRANCE	Pergamon Press SARL, 24 rue des Ecoles, 75240 Paris, Cedex 05, France
FEDERAL REPUBLIC OF GERMANY	Pergamon Press GmbH, Hammerweg 6, Postfach 1305, 6242 Kronberg/Taunus, Federal Republic of Germany

Library of Congress Cataloging in Publication Data
Main entry under title:

Mathematical biology.

Proceedings of a conference sponsored by the Molecular Science Program at Southern Illinois University at Carbondale, held May 27-28, 1980.
1. Biomathematics--Congresses. I. Burton, T.A. (Theodore Allen), 1935- . II. Molecular Science Program (Southern Illinois University at Carbondale) [DNLM: 1. Biometry--Congresses. 2. Molecular biology--Congresses. WA 950 M426 1980]
QH323.5.M3637 1981 574'.01'51 81-1352
ISBN 0-08-026348-8 AACR2

Printed in the United States of America

CONTENTS

PREFACE

On May 27-28, 1980, a conference on mathematical biology was held at Southern Illinois University in Carbondale, Illinois. The program featured distinguished scientists from across the United States as well as Belgium and Japan. The present volume consists of the proceedings of that conference.

The conference, and these proceedings, are unusual in several ways. The conference did not have a single theme. Invitations were given to outstanding investigators in fairly diverse areas of mathematical biology to talk about their own investigations at some depth. Thus, the number of speakers was fairly small and their talks were of moderate length. Each speaker was asked to submit a substantial paper for these proceedings which would provide a good introduction to their subject, present results on the frontiers, and references to related investigations. I am well pleased with the results and believe that some of these papers will do much to stimulate further investigations as well as inform the scientific community of some of the exciting work being undertaken in broad areas of mathematical biology.

The conference itself was the first in a series to be sponsored by the Molecular Science Program at SIU. In October of 1979 the program Chairman, Dr. G. V. Smith of the SIU Chemistry Department, called a meeting of the nonexperimentalists within the interdisciplinary program and suggested that an annual conference be held at SIU on some molecular science topic of interest to the group. It was decided that mathematical biology should be the first topic and an organizing committee of Professors S. Bose (Physics), M. Emptage (Chemistry), and myself (Mathematics) was chosen. The selection of speakers and the arrangement of the program was a joint effort of those committee members.

The main funding was provided through the budget of the Molecular Science Program, with substantial help from the Graduate School, the College of Liberal Arts, and the Departments of Chemistry, Mathematics, and Physics. All of that assistance is gratefully acknowledged.

I am especially indebted to Ms. Sharon Champion for the excellent typing of the papers. Thanks are also extended to the Dean of the College of Liberal Arts, Dr. James Light, for continued financial assistance with the project.

Finally, I wish to thank Pergamon Press for their interest in and support of the project.

NONLINEAR DIFFERENTIAL EQUATION MODELS IN ENZYME REGULATED PATHWAYS

H. T. Banks†
Lefschetz Center for Dynamical Systems, Division of Applied Mathematics, Brown University, Providence, Rhode Island

ABSTRACT

A summary of our recent and continuing efforts in two areas of biology is presented. First, we discuss our latest investigations of nonlinear stiff systems of coupled ordinary differential and algebraic equations which model the enzyme cascade in glycogenolysis. We then summarize our findings for nonlinear ordinary and functional differential equation models to describe control of protein synthesis.

INTRODUCTION

In this note we give a brief summary of two projects on which we have worked during the past several years. The first project summarized entails our long term cooperative and continuing efforts with R. Miech of the Division of Biomedical Sciences, and recent findings are reported in [5]. (An earlier detailed summary along with complete references can be found in Chapter 3 of [1].) These investigations concern the development of mathematical models for enzyme cascades such as that found in glycogenolysis (the hydrolysis of glycogen--a storage form of carbohydrates). The studies lead to qualitative questions for stimulus-response mechanisms in the context of nonlinear stiff systems of coupled algebraic and ordinary differential equations.

The second project we describe involves the use of mathematical models to investigate qualitative behavior of control systems which have been proposed for regulation of protein synthesis. These control systems have been studied by numerous authors using nonlinear equations of several types: ordinary, functional, and partial differential equations. Questions of paramount interest have been global asymptotic stability of equilibria and instability with existence of large amplitude sustained oscillations. The efforts which we summarize represent our joint endeavors with J. Mahaffy and are explained in more detail in [2], [3], [4], [6].

---

† This research was supported by the National Science Foundation under grant MCS-79-05774.

MODELS FOR RECYCLING ENZYME CASCADES

Enzyme cascades such as those of interest to us here are common in metabolic pathways where one has a sequence of enzyme-catalyzed reactions as illustrated in Figure 1.1.

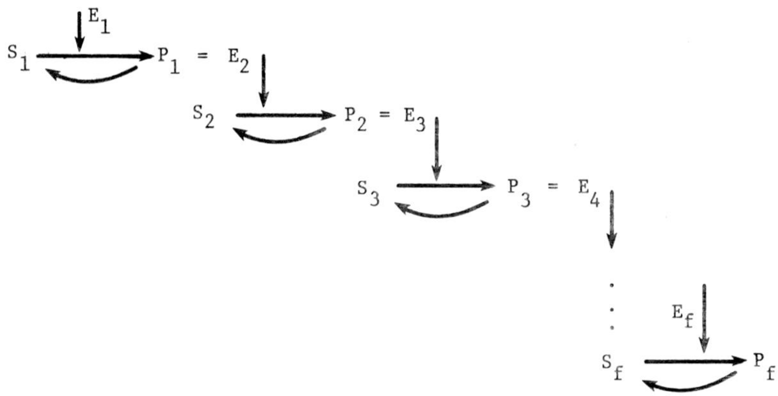

Fig. 1.1.

In addition to the sequence as depicted, one may also have feedforward and/or feedback mechanisms playing an important role in these cascades, which, through their functioning, lower energy requirements via use of a pathway between a given biochemical "stimulus" and "response." Recycling cascades as given in the schematic are found in such diverse pathways as blood coagulation, glycogenolysis, and activation of the complement system in immune response. They have been hypothesized to exist in visual excitation and in relaxation-contractile mechanisms in smooth muscle. A common feature of these cascades is that they entail a biochemical amplification system with fast response time (i.e., a small signal produces a large response very quickly).

Our own efforts have involved development of mathematical models for a cascade system to describe glycogenolysis. The underlying biochemical model on which several of our mathematical models are found is sketched in Figure 1.2 (the ' denotes the inactive form of an enzyme). We have, since 1973, developed a number of mathematical models for this cascade. These models, which have been formulated in conjunction with experimental findings, have been used to test a number of hypotheses concerning possible mechanisms of importance in the glycogenolytic pathway.

The general form for our mathematical models, based on conservation of mass, is given by a set of differential equations (see Figure 1.2)

$$\begin{aligned}
\dot{c} &= v_1(AC,\ ATP) - v_2(PDE,\ c)\\
\dot{\alpha} &= v_3(c,\ A - \alpha) - v_4(sub{:}c,\ \alpha)\\
\dot{\beta} &= v_5(\alpha,\ B - \beta) - v_6(PKP,\ \beta)\\
\dot{\gamma} &= v_7(\beta,\ G - \gamma,\ Ca) - v_8(PP,\ \gamma,\ Ca)\\
\dot{Ca} &= f(c).
\end{aligned} \tag{1.1}$$

In this system c and Ca represent the concentrations of cyclic AMP and free calcium, respectively, while α, β and γ represent the concentrations of active forms of protein kinase, phosphorylase kinase and glycogen phosphorylase, respectively. The quantities A, B, G (assumed constant) represent the concentrations of the total amount (active plus inactive form) of these same entities.

A recent version of one of these models is described in detail in [5] and is based on the assumption that activation of protein kinase (PKK' $\rightarrow$ PKK in Figure 1.2) involves third order kinetics. Using the laws of mass action, certain equilibrium assumptions, and Michaelis-Menten, Briggs-Haldane type reaction velocity approximations, the following system of equations can be derived.

$$\begin{aligned}
\dot{c}(t) &= k_c - k_d c(t)\\
\alpha^3(t) &- K_{eq}\ 2\{A - \alpha(t)\}c^2(t) = 0\\
\dot{\beta}(t) &= \frac{TN_5\alpha(t)\{B - \beta(t)\}}{K_5 + B - \beta(t) + \alpha(t)} - \frac{TN_6[PKP]\beta(t)}{K_6 + [PKP] + \beta(t)}\\
\dot{\gamma}(t) &= \frac{TN_7\beta(t)\{G - \gamma(t)\}}{K_7 + G - \gamma(t) + \beta(t)} - \frac{TN_8[PP]\gamma(t)}{K_8\left\{1 + \frac{Ca(t)}{K_E}\right\} + [PP] + \gamma(t)}\\
Ca(t) &= kc(t).
\end{aligned} \tag{1.2}$$

With appropriate values for the kinetic parameters (i.e., k_c, k_d, K_{eq}, K_E, k, the turnover numbers TN_i, the Michaelis constants K_i), (1.2) constitutes a stiff system of coupled nonlinear ordinary differential and algebraic equations. This necessitates that any investigation consist of a combination of partial analytical analysis and numerical studies employing methods (such as that due to Gear) for stiff equations.

We note that the second equation in (1.2) is actually an algebraic one resulting from a quasi-equilibrium or zero equilibrium time-assumption for the protein kinase

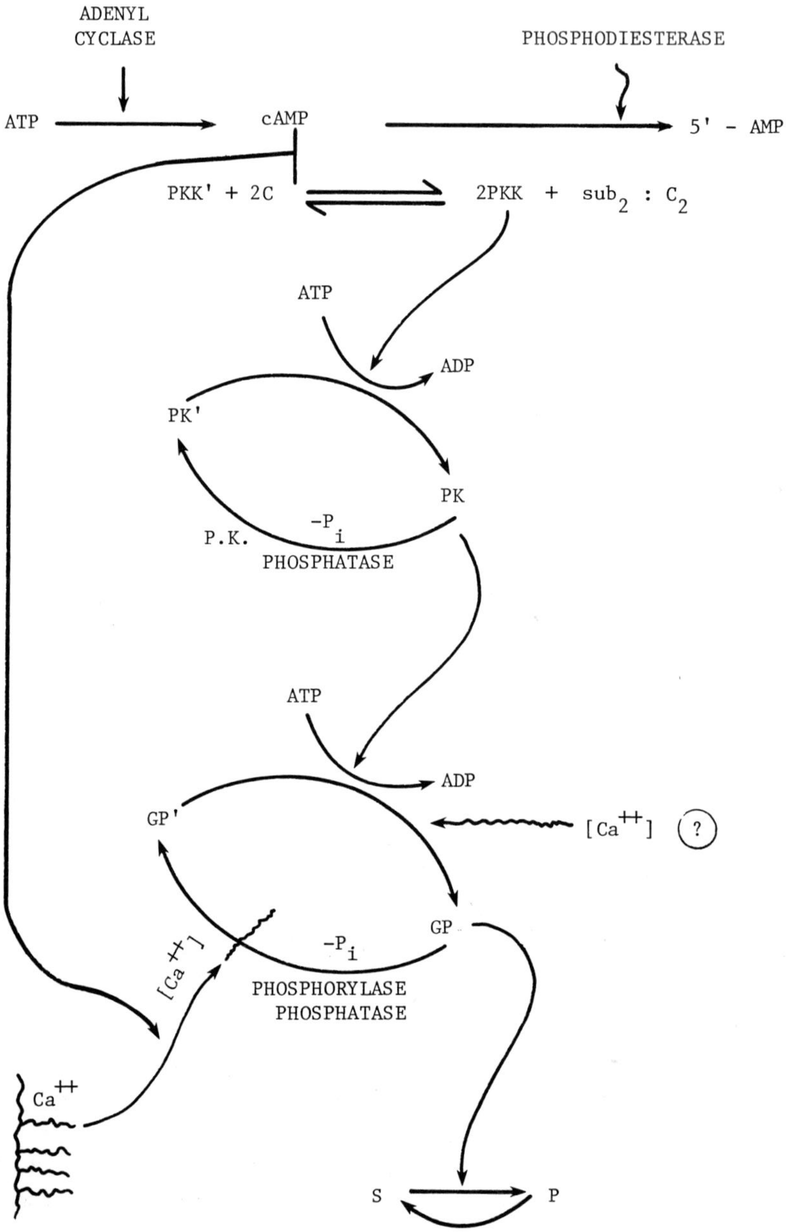
ADENYL
CYCLASE
PHOSPHODIESTERASE
ATP
cAMP
5' - AMP
PKK' + 2C
2PKK + sub2 : C2
ATP
ADP
PK'
PK
-Pi
P.K.
PHOSPHATASE
ATP
ADP
GP'
[Ca++]
?
GP
-Pi
[Ca++]
PHOSPHORYLASE
PHOSPHATASE
Ca++
S
P

Fig. 1.2.

step in the cascade. That is, in the model we have assumed $\frac{d\alpha}{dt} \approx 0$ to obtain this algebraic equation. This zero equilibrium time assumption can be explained mathematically in the following way. Let t be the time scale (say, in seconds) appropriate for observing the dynamics of c-AMP, phosphorylase kinase and glycogen phosphorylase and let τ represent that time scale suitable for observing with accuracy the changes in protein kinase in the cascade. Thus, $\frac{d\alpha}{d\tau}$ is an easily observed non-zero rate of change of α. If the dynamics for α are fast (say, on the order of micro-sec.) compared to those for c, β, γ, then we have $\tau = \varepsilon t$ where $\varepsilon \sim 10^{-6}$. It follows that

$$\frac{d\alpha}{d\tau} = \frac{1}{\varepsilon}\frac{d\alpha}{dt} \quad \text{or} \quad \frac{d\alpha}{dt} = \varepsilon \frac{d\alpha}{d\tau} \sim 10^{-6}\,\frac{d\alpha}{d\tau} .$$

Therefore, even though $\frac{d\alpha}{d\tau}$ might be an observable, measurable quantity, $\frac{d\alpha}{dt}$ will be orders of magnitude smaller and hence might reasonably be approximated by $\frac{d\alpha}{dt} \approx 0$ in a model for the overall cascade. We hasten to add that this is not a factor in the stiffness of the model (1.2) (see Chapter 3 of [1] for a discussion of this aspect of the models).

As we have already noted, the second equation in (1.2) is a result of the zero equilibrium time assumption in the differential equation

$$\dot{\alpha} = k_{+1}(A - \alpha)c^2 - k_{-1}\alpha^2\left(\frac{1}{2}\,\alpha\right),$$

which is, in turn, based upon assuming that a molecule of inactive protein kinase consists of two regulatory subunits and two catalytic subunits. Early models (see [1] and [5]) were derived under the assumption that inactive protein kinase consists of one regulatory subunit and one catalytic subunit. While current biochemical evidence tends to support the assumption that inactive protein kinase is a tetramer, our comparative studies of the two mathematical models indicate that they yield qualitatively and quantitatively almost the same type solutions. For example, the older models with initial values (in Molars) $c_0 = .1 \times 10^{-8}$, $\alpha_0 = .031 \times 10^{-7}$, $\beta_0 = .327 \times 10^{-6}$ and $\gamma_0 = .134 \times 10^{-5}$ reach steady states (after a 20-fold stimulation in c-AMP) of $c_{ss} = 2 \times 10^{-8}$, $\alpha_{ss} = .131 \times 10^{-7}$, $\beta_{ss} = .672 \times 10^{-6}$, and $\gamma_{ss} = .865 \times 10^{-5}$ while the model (1.2) detailed above with similar values for c_0 and c_{ss} possesses solutions corresponding to $\alpha_0 = .0259 \times 10^{-7}$, $\beta_0 = .288 \times 10^{-6}$, $\gamma_0 = .121 \times 10^{-5}$ with steady states given by $\alpha_{ss} = .180 \times 10^{-7}$, $\beta_{ss} = .738 \times 10^{-6}$ and $\gamma_{ss} = .875 \times 10^{-5}$. The trajectories for both models are so similar as to be essentially indistinguishable from a qualitative viewpoint.

As reported in earlier discussions ([1], [5]), it is as yet unknown whether the increased free calcium (due to increased levels of cAMP) feeds forward in the cascade to inhibit the reaction GP → GP' (this is the mechanism implicit in the equations (1.2) above) or whether it acts as an activator in the reaction GP' → GP. In the latter case one might take

$$v_7 = \frac{\widetilde{TN}_7 Ca(t)\beta(t)\{G - \gamma(t)\}}{\tilde{K}_7 + G - \gamma(t) + \beta(t)}$$

and

$$v_8 = \frac{\widetilde{TN}_8 [PP]\gamma(t)}{\tilde{K}_8 + [PP] + \gamma(t)}$$

in the model equations (1.1). Our studies comparing such a model with (1.2) reveal that there is essentially no difference qualitatively or quantitatively between the characteristic features (amplification, fast response time, etc.) of these models.

Our studies over the past several years have aided in the understanding of the biochemistry of glycogenolysis and the roles certain mechanisms (e.g., autocatalytic, cAMP induced Ca^{++} release) play in the amplification and fast response features of the cascade (see [1] and [5] for more details). From a mathematical viewpoint, we understand a great deal more about the general techniques (both qualitative and numerical) that must be used in the formulation and evaluation of such models.

While our efforts on glycogenolysis models are continuing, we are now beginning to apply our general approach to develop models for the hypothesized cascade in bronchial smooth muscle with a long term goal of use of the resulting models and the ensuing knowledge gained from these efforts to aid in drug design and development.

MODELS FOR PROTEIN SYNTHESIS

We recall that proteins consist of one or more polypeptide chains which are themselves nothing more than hundreds of amino acids linked together in specific sequences. There are many types of proteins, but the largest class are enzymes (of utmost importance because of their catalytic function in cells). The code (genetic code) or template for construction of the ordered sequence in a polypeptide is found in the nucleic acids DNA (deoxyribonucleic acid) and RNA (ribonucleic acid) which are chainlike macromolecules which function in the storage and transfer of genetic information.

With regard to regulation of the synthesis of proteins, we observe that the level of understanding of the underlying mechanisms and control differ greatly between prokaryotic (simple organisms such as bacteria with no well-defined cellular nucleus surrounded by a membrane) and eukaryotic (higher organisms such as man in which cells contain a nuclear membrane) organisms. In the latter, regulation of protein synthesis and, indeed, the pathways themselves are still a matter of uncertainty and extensive biological research. In the case of prokaryotic cells, there do exist accepted biological models (the so-called Jacob-Monod schemes) for the control of gene expression. There are two basic models, similar in spirit: induction (positive transcriptional control) and repression (negative transcriptional control). In both schemes one has structural genes which carry the information for coding the specific amino acid sequence of a particular polypeptide, an operator gene which controls the activity of the structural genes, and a regulatory gene which codes for a repressor (which may be active or inactive, depending on the specific pathway) of the operator gene. Control in both models is at the level of transcription (the process whereby information in the structural genes is encoded in a template, messenger RNA, which is used at the ribosomes in a process--translation--resulting in production of polypeptide chains). That is, mRNA is not made unless it is to be used to produce protein. Since we shall be concerned here only with the repression scheme, we shall not discuss the induction model further but instead refer the reader to [4] for a complete presentation of the details of the induction.

Turning to repression schemes, we note that in the classical Jacob-Monod scheme as depicted in Figure 1.3, the aporepressor is inactive in its free form and must first bind with an endproduct (corepressor) to form the active repressor complex which is capable of binding and repressing the operator gene. The biosynthesis of

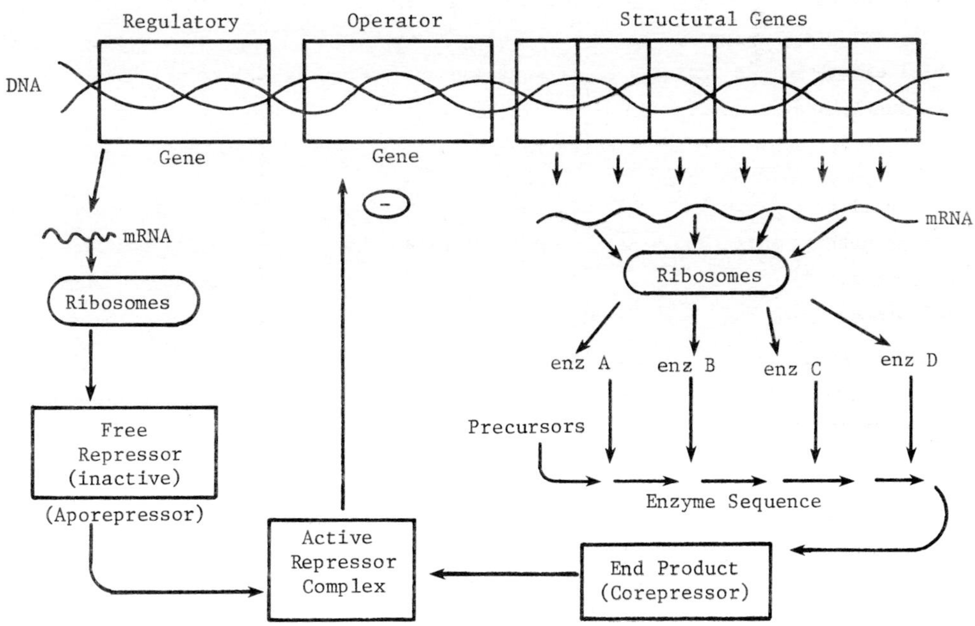

Fig. 1.3.

tryptophan, an amino acid, is known to follow a classical repression scheme with five structural genes coding for five enzymes which catalyze a sequence of reactions with tryptophan the endproduct which serves as a corepressor. Other biosynthetic pathways, such as that for histidine (another amino acid), often follow a modified Jacob-Monod scheme.

Mathematical models for endproduct repression have been the subject of investigations for a number of years now, dating back at least to the early models of B. Goodwin (for a detailed summary of the literature, see [4]), and some are based on the simple schematic given in Figure 1.4.

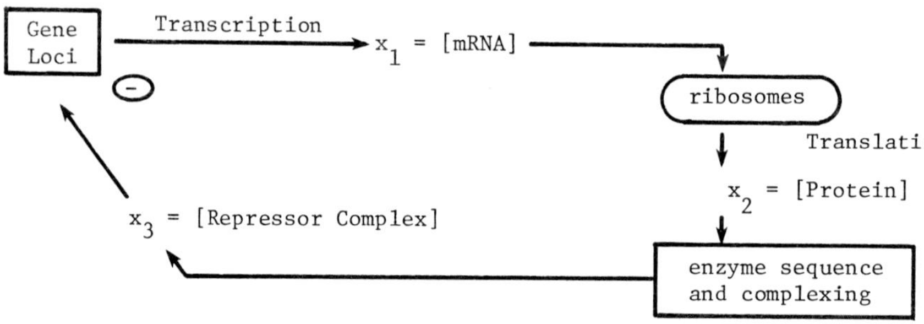

Fig. 1.4.

More sophisticated ordinary differential equation models and models involving differential equations with delays (both discrete and distributed) can be derived in a systematic way by first considering the stoichiometric equations

$$G + NTP \rightleftharpoons G\cdot NTP \longrightarrow M + G + (DP)'s$$

$$M + AA \rightleftharpoons M\cdot AA \longrightarrow M + E + tRNA$$

$$E + S \rightleftharpoons E\cdot S \longrightarrow E + P$$

$$P + AR \rightleftharpoons R$$

$$R + G \rightleftharpoons G\cdot R,$$

which are based on the schematic given in Figure 1.5. A careful derivation of the models we discuss here can be found in [4].

A central role in almost all previous mathematical investigations was played by the question of existence of sustained oscillations (i.e., periodic solutions). In some cases the effort was to establish global (in all our discussions "global" refers to the nonnegative orthant $x_i \geq 0$, $i = 1,\ldots, n$, since that is the only region in which the model equations are assumed to be well-defined) asymptotic stability of

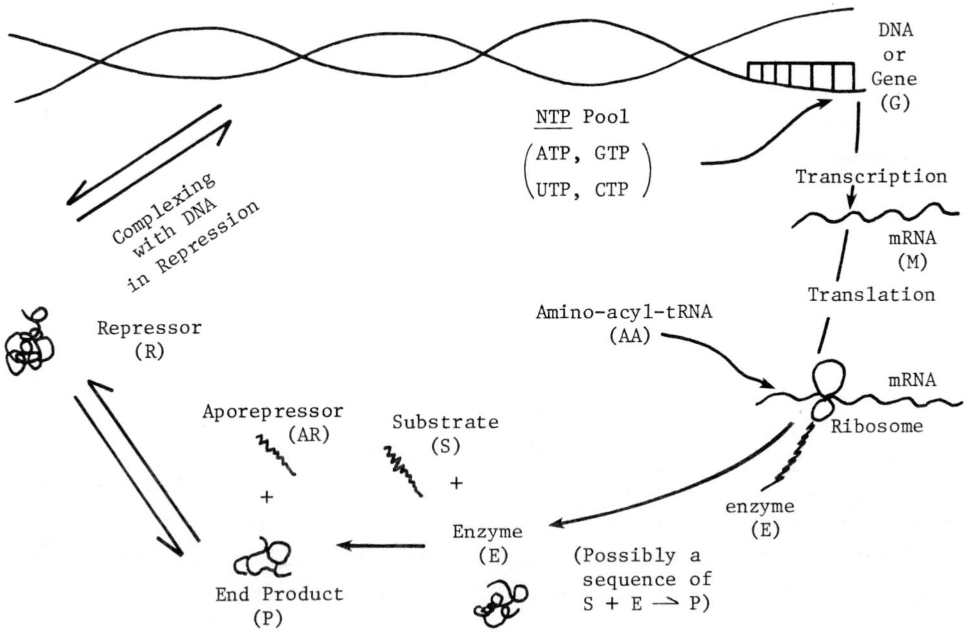

Fig. 1.5.

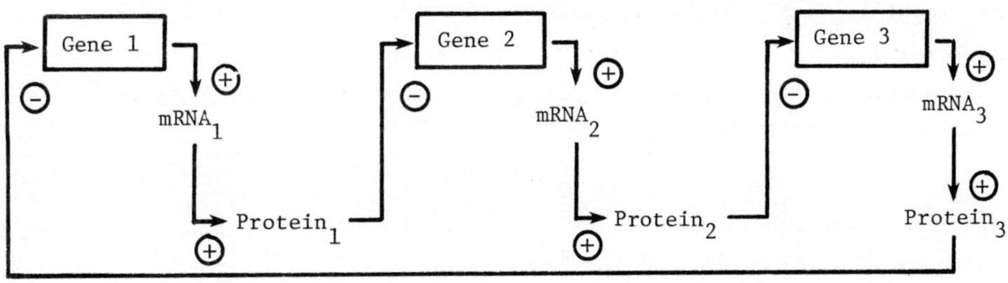

Fig. 1.6.

an equilibrium solution thereby ruling out the possibility of oscillations. This was exactly the situation in [3] where we considered cyclic gene models proposed by several authors (Goodwin; Tiwari and Fraser; and others) in previous treatments and based on a schematic such as that given in Figure 1.6 for a three-gene configuration. In [3] and [4] we considered very general cyclic gene models of which the following n-gene model is a special case:

$$\begin{aligned}
\dot{x}_1(t) &= \frac{a_1}{1 + k_1 L_1^1(z_n^t)} - b_1 x_1(t) \\
\dot{y}_1(t) &= \alpha_1 L_1^2(x_1^t) - \beta_1 y_1(t) \\
\dot{z}_1(t) &= \gamma_1 L_1^3(y_1^t) - \delta_1 z_1(t) \\
\dot{x}_i(t) &= \frac{a_i}{1 + k_i L_i^1(z_{i-1}^t)} - b_i x_i(t) \\
\dot{y}_i(t) &= \alpha_i L_i^2(x_i^t) - \beta_i y_i(t) \\
\dot{z}_i(t) &= \gamma_i L_i^3(y_i^t) - \delta_i z_i(t), \qquad i = 2, 3, \ldots, n,
\end{aligned} \tag{1.3}$$

where the L_i's are operators consisting of a finite number of saltus functions plus an absolutely continuous kernel. That is, x^t, y^t, z^t denote the functions $\theta \to x(t+\theta)$, $y(t+\theta)$, $z(t+\theta)$, respectively and the operators L_i^j are given by

$$L_i^j(z^t) = \sum_{\ell=0}^{\nu} c_{i\ell}^j \, z(t - h_\ell) + \int_{-r}^{0} z(t+\theta)\zeta_i^j(\theta)d\theta.$$

Using differential inequalities and basic properties of delay equations one can establish the following stability result.

<u>THEOREM</u>. In the orthant $P = \{(x,y,z) \in R^{3n} | x_i \geq 0,\ y_i \geq 0,\ z_i \geq 0\}$ the system (1.3) has a unique equilibrium solution $\bar{x}$, $\bar{y}$, $\bar{z}$. Given any continuous initial data ϕ with values lying in P (i.e., $\phi(\theta) \in P$, $-r \leq \theta \leq 0$), there exists a unique solution of (1.3) defined for $0 < t < \infty$ satisfying $(x(t), y(t), z(t)) \in P$ for $t \geq 0$. Furthermore as $t \to \infty$, we have $(x(t), y(t), z(t)) \to (\bar{x},\bar{y},\bar{z})$.

This result is valid independent of whether n (the number of genes in the configuration) is odd or even. This shows that the conjectures (that odd gene configurations would exhibit sustained oscillations) of Tiwari and Fraser based on computer simulation studies are not true.

Mahaffy, in his thesis and [6], established positive results for a rather general class of single gene models in which cooperative inhibition is allowed. He considered the system

$$\begin{aligned}
\dot{x}_1(t) &= \frac{a}{1 + k[L_1(x_n^t)]^\rho} - bx_1(t) \\
\dot{x}_i(t) &= \alpha_i L_i(x_{i-1}^t) - \beta_i x_i(t), \qquad i = 2, \ldots, n,
\end{aligned} \tag{1.4}$$

and established the following:

(i) the system (1.4) has a unique equilibrium solution in the positive orthant of R^n;

(ii) for $\rho = 1$, this solution is globally asymptotically stable (these results are essentially a special case of the conclusions obtained in the above theorem and are given in [2]);

(iii) for $\rho > 1$ there are two possibilities: Either (a) the system (1.4) possesses a globally asymptotically stable equilibrium (in the event a certain related $2\rho + 1$-order algebraic equation has a unique positive root) or, (b) in the event (a) does not obtain, we at least have the existence of a bounded region $\mathcal{R}$ in R^n_+ that is attracting and invariant for (1.4).

The proofs for the above results again involve only use of differential inequalities and asymptotic properties of solutions of delay equations and are very similar to the arguments found in [2]. By further restricting the system (1.4) to the case of discrete delays, Mahaffy [6] has been able to show that the region $\mathcal{R}$ described in (iii)-(b) actually contains a nonconstant periodic solution. He considered the special case

$$\begin{aligned} \dot{x}_1(t) &= \frac{a}{1 + kx_n(t - \tau_1)^\rho} - bx_1(t) \\ \dot{x}_i(t) &= \alpha_i x_{i-1}(t - \tau_i) - \beta_i x_i(t), \qquad i = 2,\ldots, n, \end{aligned} \tag{1.5}$$

and showed that, under certain conditions on τ_i, ρ, α_i, β_i, a and k, (1.4) possesses nonequilibrium periodic solutions. The proof, given in [6], is very technical (for a summary see [4]) and involves use of a fixed point theorem due to Nussbaum. Particular cases of the result yield the following:

<u>THEOREM</u>. Suppose

(i) $r = \sum_{i=1}^{n} \tau_i$ is sufficiently large so that

$$\text{Arctan}\,\frac{\pi}{rb} + \sum_{j=2}^{n-1} \text{Arctan}\,\frac{\pi}{r\beta_j} < \pi;$$

(ii) there exist $\delta > 0$ and $c > (b/\bar{\alpha})\{e^{\bar{\beta}r} - 1\}^{-1}$, where $\bar{\alpha} = \prod_{j=2}^{n} (\alpha_j/\beta_j)$ and $\bar{\beta} = \min\{b, \beta_2,\ldots, \beta_n\}$, such that $f(\xi) = -b\bar{x}_1 + \dfrac{a}{1 + k(\xi + \bar{x}_n)^\rho}$ satisfies $|f(\xi)| \geq c|\xi|$ for $|\xi| < \delta$;

(iii) the parameters τ_i, ρ, α_i, β_i, a, k are such that

$$F(\lambda) = (b + \lambda) \prod_{j=2}^{n} (\beta_j + \lambda) - \left(\prod_{j=2}^{n} \alpha_j \right) f'(0) e^{-\lambda r}$$

has zeros in the complex plane with positive real parts.

Then (1.5) has a nonconstant periodic solution.

Corollaries to this theorem include the useful statements:

(A) For n = 3, (i) above holds for all r and one only needs to establish (ii) and (iii) to ensure veracity of the conclusion of the theorem.

(B) If $b\left(\prod_{j=2}^{n} \beta_j \right) < -f'(0)\left(\prod_{j=2}^{n} \alpha_j \right)$, then for r sufficiently large (1.5) has a periodic solution.

To investigate further these nonconstant periodic solutions and to better understand qualitative dependence of the model (1.5) on the parameters, a large number of numerical simulations were carried out. These are reported in some detail in [4] and we only briefly summarize some of our findings here. Taking $\tau_3 = 0$ and $n = 3$ in (1.5), available biological data suggests that $\tau_1 + \tau_2 \approx 1$ minute is a reasonable estimate for the delay parameters, that $\rho = 2$ and possibly $\rho = 3$ are acceptable in repression models ($\rho = 4$ has been suggested only for induction pathways; in fact the degree of cooperativity ρ is not well-documented in most pathways). The parameters b and β_2 represent the half-lives of mRNA and an enzyme, respectively, while β_3 is more difficult to estimate since it is related to the rate of uptake and degradation of endproduct. For b one has that .5 min^{-1} is a reasonable choice while β_2 might range from .2 to 1 min^{-1}. Our best guess on a range for β_3 is .1 to 10^2 min^{-1}. By a rescaling of (1.5), it is clear that the parameters a, α_2 and α_3 can be chosen arbitrarily (a value of 10 was chosen for these parameters). For $k = 10$ and $b = .5$, the system (1.5) possesses a periodic solution with period of μ minutes as reported in the following table of sample numerical results.

ρ	β_2	β_3	τ_1	τ_2	μ
2	.2	.5	5	3	34
2	.2	1	3	3	25
3	.2	1	3	3	29.5
3	.2	1	2	2	21.5
4	.2	1	1	1	14
4	.5	1	.5	.5	10
6	.2	1	.5	.5	11

In general one finds that the model (1.5) is not very sensitive to the values of k, β_2 and β_3 while it is highly sensitive to the values of ρ, τ_1 and τ_2. There appears to be a critical relationship between these latter parameters for periodic solutions to exist. For $\rho = 4$, one obtains oscillations with biologically acceptable values of all other parameters while for $\rho = 2$, the values of τ_1 and τ_2 required appear to be slightly too large.

Further numerical findings and some preliminary findings for other models that involve modifications of (1.5) are reported in [4]. From our studies it appears that there are a number of interesting directions for further research on these problems.

In concluding we mention briefly a class of models that we feel merit further investigation. These models would incorporate ideas for delay-represented processing and transport phenomena such as those discussed by Thames and Aronson in [7]. The models might serve as a first attempt at modeling protein synthesis in eukaryotic cells. One assumes that there is a promoter u (e.g., mRNA) and inhibitor v (e.g., an endproduct) coupled as shown in Figure 1.7 with distinct sites (separated by a diffusion region $\mathcal{D}$) for their activity.

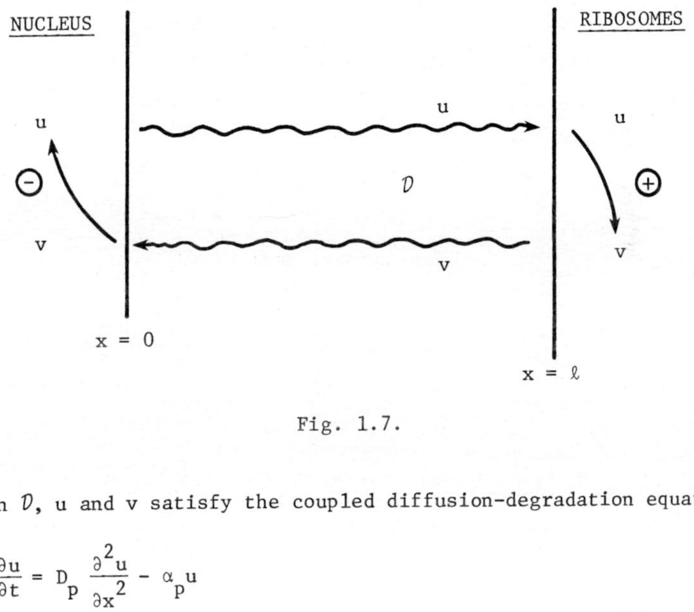

Fig. 1.7.

In the region $\mathcal{D}$, u and v satisfy the coupled diffusion-degradation equations

$$\frac{\partial u}{\partial t} = D_p \frac{\partial^2 u}{\partial x^2} - \alpha_p u$$

$$\frac{\partial v}{\partial t} = D_i \frac{\partial^2 v}{\partial x^2} - \alpha_i v. \qquad (1.6)$$

The promoter u acts as an activator for production of v at $x = \ell$ (the "RIBOSOMES") while v is an inhibitor of u production at $x = 0$ (the "NUCLEUS"). Boundary conditions might be taken as

$$\begin{aligned} u(0,t) &= V_1(v(0,t-\tau_2)) \\ \frac{\partial v}{\partial x}(0,t) &= 0 \\ \frac{\partial u}{\partial x}(\ell,t) &= 0 \\ v(\ell,t) &= V_2(u(\ell,t-\tau_1)). \end{aligned} \tag{1.7}$$

Here $u(0, t)$ is the concentration of the promoter after (i) packaging and transport of the inhibitor v across the nuclear membrane, (ii) transcription, and (iii) packaging and transport of mRNA across the nuclear membrane. The delay τ_2 is the time required for (i), (ii) and (iii). At the other boundary, $v(\ell, t)$ is the concentration of the endproduct after a delay time τ_1 for translation of mRNA and any enzymatic reaction sequence required to produce the endproduct repressor. The nonlinear functions V_1 and V_2 are reaction velocity approximations for inhibition and activation respectively (e.g., $V_1(z) \sim \frac{1}{1+z^\rho}$, $V_2(z) \sim \frac{z}{1+z}$).

Given the results for delay systems outlined above and those for separated cooperative enzyme systems in [7], one might reasonably expect to find that the models (1.6), (1.7) exhibit sustained oscillations for delay times τ_1, τ_2 and separation distances ℓ that are consistent with eukaryotic cell structure and biochemistry.

REFERENCES:

1. Banks, H. T., Modeling and Control in the Biomedical Sciences, Lecture Notes in Biomathematics, Vol. 6, Springer-Verlag, Berlin, 1975.

2. Banks, H. T. and Mahaffy, J. M., Global asymptotic stability of certain models for protein synthesis and repression, Quart. Appl. Math. 36: 209-221 (1978).

3. Banks, H. T. and Mahaffy, J. M., Stability of cyclic gene models for systems involving repression, J. Theor. Biol. 74: 323-334 (1978).

4. Banks, H. T. and Mahaffy, J. M., Mathematical Models for Protein Synthesis, LCDS Tech. Rep. 79-4, Brown University, Providence, August, 1979.

5. Banks, H. T., Miech, R. P. and Olson, S. L., A comparison of mathematical models for a recycling cascade in glycogenolysis, Intl. J. Math. Modeling, to appear.

6. Mahaffy, J. M., Periodic solutions for certain protein synthesis models, J. Math. Anal. Appl. 74: 72-105 (1980).

7. Thames, H. D. and Aronson, D. G., Oscillation in a nonlinear parabolic model of separated, cooperatively coupled enzymes, in Nonlinear Systems and Applications (V. Lakshmikantham, ed.), Academic Press, New York, 1977, 687-693.

EXISTENCE OF A PERIODIC SOLUTION FOR A TWO PREDATOR-ONE PREY ECOSYSTEM MODELED ON A CHEMOSTAT

G. J. Butler†
University of Alberta, Edmonton, Alberta and
University of Iowa, Iowa City, Iowa

Paul Waltman††
University of Iowa, Iowa City, Iowa

ABSTRACT

A three dimensional system of ordinary differential equations modeling two predators competing for a renewable resource is analyzed and a periodic solution established in the open octant by a bifurcation from a two dimensional limit cycle. The basic equations are those of the chemostat. This result confirms numerical evidence from [5], partially answering a question raised there.

INTRODUCTION

The chemostat is a piece of laboratory apparatus frequently used for studying competition between microorganisms. The equations describing the chemostat can be analyzed mathematically [3], [6] and laboratory experiments based on these results have been performed [2]. If in the chemostat, instead of a single input nutrient, the basic resource is a renewable one, the input concentration in the equations can be replaced by a logistic term. The equations then take the form (for two competitors)

$$s' = \gamma s\left(1 - \frac{s}{K}\right) - \frac{m_1}{y_1}\frac{sx_1}{a_1 + s} - \frac{m_2}{y_2}\frac{sx_2}{a_2 + s}$$

$$x_i' = \frac{m_i s x_i}{a_i + s} - D_i x_i, \qquad i = 1, 2 \tag{1}$$

$$x_i(0) = x_{i0} \geq 0, \qquad s(0) = s_0 > 0.$$

In (1), γ is the growth rate and K the carrying capacity for the prey s while y_i, m_i, D_i, a_i are the yield constant, the intrinsic growth rate, the death rate and the Michaelis-Menten constant, respectively, for the ith predator x_i, $i = 1, 2$. The nonlinear functional response in (1) is called Michaelis-Menten Dynamics (for

---

† Research supported by NSERC grant A-8130.

††Research supported by NSF grant MCS-7901069.

enzyme reactions) or a Hollings response (for ecological problems). For a general survey of mathematical results on competition in a chemostat, see [10].

The system (1) was partially analyzed in [4] and some numerical studies and interpretations are given in [5]. The simulations in [5] posed a number of open mathematical questions (see also [7]). This paper addresses two of these which concern the existence of a periodic solution in the open positive octant.

The relevant background from [5], and some results obtained since that paper appeared are summarized in the next section and the principal theorems are stated in the final section. The proofs of the theorems will appear elsewhere.

BACKGROUND

As in [11], if one uses x_i/y_i as a variable, then (1) is obtained with $y_1 = y_2 = 1$, that is, we consider

$$s' = \gamma s\left(1 - \frac{s}{K}\right) - \frac{m_1 s x_1}{a_1 + s} - \frac{m_2 s x_2}{a_2 + s}$$

$$x_i' = \frac{m_i s x_i}{a_i + s} - D_i x_i, \qquad i = 1, 2 \tag{2}$$

$$x_i(0) = x_0 \geq 0, \qquad s(0) = s_0 > 0.$$

One can also scale out one of γ, m_1, or m_2 but we choose not to do this. Let $\lambda_i = a_i D_i/(m_i - D_i)$ (when $m_i > D_i$), $b_i = m_i/D_i$, and adopt the convention that the competitors are labeled so that $\lambda_1 < \lambda_2$. In [4], the asymptotic behavior of the system (2) was analyzed for all values of the parameters except the case $a_1 < a_2$, $1 < b_1 < b_2$. In this case with $K > a_1 + 2\lambda_1$, there was numerical evidence of coexistence in the form of a periodic solution in the open positive octant, a periodic solution which bifurcated from the limit cycle in the $s - x_1$ plane and which eventually collapsed into the limit cycle in the $s - x_2$ plane. The first conditions for coexistence--but not the existence of a periodic solution--were given in [1]. There was also a coexistence result in [11] for the special case $\lambda_1 = \lambda_2$. The first existence theorem for a periodic solution was established in [9] for $\lambda_2 - \lambda_1$ small. In this case stability was also established. Our results show that for a certain range of parameters a bifurcation occurs of the two dimensional limit cycle in the $s - x_1$ plane into the positive octant. Further we show that for K large, a periodic solution collapses into the $s - x_2$ plane.

To better set the problem it is necessary to consider the two dimensional system

$$s' = \gamma s\left(1 - \frac{s}{K}\right) - \frac{mxs}{a + s}$$

$$x' = \frac{mxs}{a + s} - Dx \tag{3}$$

$$x(0) > 0, \qquad s(0) > 0.$$

For $K > \lambda$, $\lambda = \frac{aD}{m - D}$, (3) has a unique critical point $(\hat{s}, \hat{x})$ in the positive quadrant, $\hat{s} = \lambda$, $\hat{x} = [(K - \lambda)(a + \lambda)]/K$. It was shown in [4] that, for $K < a + 2\lambda$, (3) has no periodic solutions and that for $K > a + 2\lambda$, $b > 1$, (3) possessed at least one limit cycle. The former was shown to follow from the Dulac Criterion and the latter from the Poincaré-Bendixon Theorem.

PRINCIPAL RESULTS

It was observed in [10] that the limit cycle for (3) arises from a Hopf bifurcation. The necessary stability conditions for this were established in [9]. For convenience we state this as a preliminary result to connect the critical point and the periodic solution.

THEOREM [9]. As K increases through $K_0 = a + 2\lambda$, the critical point $(\hat{s}, \hat{x})$ of (3) experiences a Hopf bifurcation in which the critical point becomes unstable, giving rise to a locally unique asymptotically stable periodic orbit in the positive quadrant for the values of K greater than but near to K_0.

The Hopf theorem yields only local uniqueness of the limit cycle but, in fact, we will be able to establish global uniqueness for certain ranges of values of the parameter K. We are led then to make the following basic assumption.

HYPOTHESIS H. (3) has a unique periodic orbit lying in the positive quadrant. One of the Floquet exponents of the linearization about this orbit has negative real part.

We note that (H) implies that the periodic orbit is asymptotically orbitally stable.

Sufficient conditions for (H) to hold will be given below. There is strong numerical evidence for uniqueness and asymptotic (orbital) stability of the periodic orbit, for all $K > K_0$.

The first theorem yields the existence of a periodic solution in the positive octant for (2) arising from the two dimensional limit cycle.

THEOREM 1. Let a_1, m_1, D_1, K be given with $K > a_1 + 2\lambda_1$, and assume that hypothesis (H) holds for the system (2) with $m = m_1$, $a = a_1$, $D = D_1$. Let

$a_2 > a_1$ and m_2 be given. Then $D_2 < D_1$ may be chosen so that (2) has a periodic orbit in the positive octant, arbitrarily near the $s - x_1$ plane.

The proof proceeds via a bifurcation theorem [8, p. 24] applied to a Poincaré map in the neighborhood of the two dimensional limit cycle.

The collapse of the three dimensional limit cycle for (2) into a two dimensional limit cycle for large K, is the equivalent to a bifurcation from a two dimensional limit cycle. That this occurs for some parameter range is contained in the next result.

THEOREM 2. Let a_2, m_2, D_2 be given and assume that hypothesis (H) holds for the system (2) with $m = m_2$, $a = a_2$, $D = D_2$, and for K sufficiently large. Then, given m_1 and $a_1 < a_2$, D_1 may be chosen with $\lambda_1 \leq \lambda_2$ so that (2) has a periodic orbit in the positive octant arbitrarily near the $s - x_2$ plane.

This theorem depends on the following technical lemma.

LEMMA. Let a_1, a_2, m_2, D_2 be given and let $\varepsilon > 0$. Then there exists a $K_2 = K_2(a_1, a_2, m_2, D_2, \varepsilon)$ such that if $K > K_2$ and if hypothesis (H) holds then on defining

$$\rho(a) = \frac{1}{T}\int_0^t \frac{s(t)}{a + s(t)}\, dt$$

where $(s(t), x(t))$ is the unique periodic solution of (3) of period T for $a = a_2$, $m = m_2$, $D = D_2$, we shall have $\rho(a) - D_2 < \varepsilon$ for $a_1 \leq a \leq a_2$.

It remains to provide sufficient conditions for (H) to hold. This is given for two sets of circumstances.

THEOREM 3. Let a, m and D be given. Then there exists $\varepsilon > 0$, $K_1 > a + 2\lambda = K_0$, such that Hypothesis (H) holds:

(i) if $K_0 < K < K_0 + \varepsilon$

or

(ii) if $K > K_1$.

The major ingredient for (ii) is an asymptotic estimate in K of the time spent in traversing an orbit.

REFERENCES:

1. Butler, G. J., Coexistence in Predator-Prey Systems, in Modeling and Differential Equations in Biology, T. Burton, editor, Marcel Dekker, New York, in press.

2. Hansen, S. R. and Hubbell, S. P., Single-nutrient microbial competition: Agreement between experimental and theoretical forecast outcomes, Science, to appear.

3. Hsu, S. B., Hubbell, S. and Waltman, P., A mathematical theory for single-nutrient competition in continuous cultures of microorganisms, SIAM J. Appl. Math. 32: 366 (1977).

4. Hsu, S. B., Hubbell, S. P. and Waltman, Paul, Competing predators, SIAM J. Appl. Math. 35: 617 (1978).

5. Hsu, S. B., Hubbell, S. P. and Waltman, Paul, A contribution to the theory of competing predators, Ecological Monographs 48: 337 (1978).

6. Hsu, S. B., Limiting behavior for competing species, SIAM J. Appl. Math. 34: 760 (1978).

7. Koch, A. L., Competitive coexistence of two predators utilizing the same prey under constant environmental conditions, J. Theor. Biol. 44: 378 (1974).

8. Marsden, J. E. and McCracken, M., The Hopf Bifurcation and Its Applications, Springer-Verlag, New York, 1976.

9. Smith, H. L., Coexistence of Two Competing Predators. Preprint, 30 pp.

10. Waltman, P., Hubbell, S. P. and Hsu, S. B., Theoretical and experimental investigations of microbial competition in continuous culture, in Modeling and Differential Equations in Biology, T. Burton, editor, Marcel Dekker, New York, in press.

11. Wilken, D. R., Some remarks on a competing predator problem. Preprint, 16 pp.

ON GROWTH AND FORM

J. Chadam, Department of Mathematics, Indiana University, Bloomington, Indiana

P. Ortoleva, Departments of Chemistry and Geology, Indiana University, Bloomington, Indiana

ABSTRACT

Spontaneous symmetry breaking associated with amorphous solid growth is treated analytically. The theory is quite general, avoiding the steady state diffusion and equilibrium surface concentration approximations. The stability problem is shown to be analytically soluble in terms of Whittaker functions of a similarity variable.

INTRODUCTION

Growth in Nature is often accompanied by spontaneous transitions to modes which possess less and less symmetry. Classic examples can be found in almost all disciplines (e.g., coral growth, snowflakes). One imagines that the omni-present random fluctuations or imperfections cause modes of less symmetry to be spontaneously initiated and subsequently grow from more symmetric but unstable modes.

In this paper we examine the above phenomenon in the context of the growth of an amorphous solid from a diffusing growth material which surrounds it. The macroscopic equations which define our model are derived in the second section. Our model generalizes in two respects the now classic model of Mullins and Sekerka ([8], [9]) (c.f., in addition, the excellent review article of Langer [7] and the references therein). We do not impose the steady state diffusion assumption nor do we insist that surface attachment kinetics are so rapid as to constrain the surface concentrations to their equilibrium values. The resulting mathematical problem is a generalization of the classic Stefan problem ([10], [6; Chap. 8]). In the third section an explicit spherically symmetric solution is found for the fully non-linear problem. The linear first variational equations with respect to this spherical solution are given in the last section and a type of linear stability analysis is carried out. It differs from the usual stability analysis in that there is no bifurcation parameter which governs the stable-unstable transitions nor is the decay-growth rate exponential. In fact we find that bumps associated with the spherical harmonic $Y_{\ell m}$ typically have an asymptotic $t^{(\ell-1)/2}$ growth rate for all $\ell \geq 0$ and $|m| \leq \ell$. For $\ell \geq 2$ this is done by explicitly finding many solutions (i.e., initial data) which possess this growth rate. Our results differ from those derivable for the Mullins and Sekerka model in two important respects. First, we find "finite mass" solutions (i.e., the perturbed concentrations are integrable outside the growing body for all time) for all harmonics $\ell \geq 2$ while the steady state solutions satisfy this criterion only for $\ell \geq 3$. Furthermore, this physically relevant class of finite mass solutions is so large in this less restrictive model that the notion of a "critical radius," R_ℓ at which the $Y_{\ell m}$ instability occurs, does not appear here as simply as it does in the steady state approximation.

We suggest that the explicit results presented here are typical of the generic results one expects from this problem. In this summary we have attempted to give enough of the details so as to make the exposition mathematically self-contained but not to obscure the simplicity of the model and results. Complete details are available elsewhere ([3], [4]).

FORMULATION OF THE PROBLEM

Consider a particle growing under isothermal conditions from an infinite medium. The surface of the body of interest is given by $S(\underline{x}, t) = 0$ while the medium interacting with the body and from which it grows is taken to be on the outside, $S > 0$ (see Figure 3.1).

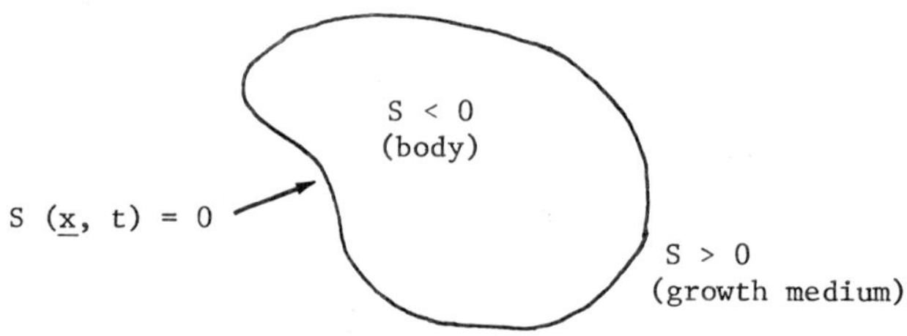

Fig. 3.1.

We consider a single concentration in growth medium which is allowed to diffuse but not react; i.e., c is a non-negative-valued function on $S > 0$ which satisfies

$$\frac{\partial c}{\partial t} = D\Delta c, \quad \text{on} \quad S(\underline{x}, t) > 0. \tag{3.1a}$$

Inside, $S < 0$, the body is taken to be a solid of constant density ρ. The jump conditions across the boundary can thus be formulated as one-sided boundary conditions relative to $S > 0$.

To obtain the kinematical (Stefan) condition, differentiating $S(\underline{x}, t) = 0$ with respect to t, at $S = 0$, gives $\frac{\partial S}{\partial t} + \dot{\underline{x}} \cdot \nabla S = 0$, on $S(\underline{x}, t) = 0$. The velocity of advancement of the interface due to growth is taken to be a local function of the concentration on the boundary and, because of surface tension effects, of the mean curvature of the surface, $K = (2|\nabla S|^3)^{-1}(|\nabla S|^2 \Delta S - \frac{1}{2}\nabla(|\nabla S|^2) \cdot \nabla S)$. That is $\dot{\underline{x}} = \underline{v}(c, K)$ where $\underline{v}$ is a phenomenological function determined from geo-chemical, physiological, etc. considerations. To further simplify the analysis we shall only consider normal growth; i.e., $\underline{v}(c, K) = v(c, K)\underline{n} = v(c, K)(|\nabla S|)^{-1}\nabla S$. Thus the kinematical condition reduces to

$$\frac{\partial S}{\partial t} + v(c, K)|\nabla S| = 0, \qquad \text{on } S = 0. \tag{3.1b}$$

In addition the flux must balance on the boundary;

$$D\nabla c \cdot \frac{\nabla S}{|\nabla S|} = (\rho - c)v(c, K), \qquad \text{on } S = 0. \tag{3.1c}$$

To complete the theory, we prescribe initial data and the ambient concentration;

$$c(\underline{x}, t_0) \text{ given on } S(\underline{x}, t_0) > 0, \tag{3.1d}$$

$$c(\underline{x}, t) \to c_\infty \text{ (constant) as } |x| \to \infty.$$

Here, we shall treat the simple model in which velocity is proportional to the difference between the concentration in the medium just at the surface and the curvature dependent equilibrium concentration; i.e.,

$$v(c, K) = \frac{k}{\rho}[c(\underline{x}, t) - c_{eq}(1 + \Gamma K)] \tag{3.2}$$

where k is a constant, c_{eq} is the equilibrium concentration for a planar interface and Γ is proportional to the surface tension.

In order to understand which are the relevant parameters in the problem, it will be convenient to non-dimensionalize Equations (3.1) with (3.2). The constant k in (3.2) has dimensions of velocity so that D/k and D/k^2 have units of length and time respectively. Making the change of variables $\underline{x}' = (k/D)\underline{x}$, $t' = (k^2/D)t$ and $c' = c/\rho$, the equations become

$$\frac{\partial c'}{\partial t'} = \Delta' c' \qquad S > 0 \tag{3.3a}$$

$$\left.\begin{aligned} &\frac{\partial S}{\partial t'} + |\nabla' S|(c' - c'_{eq}(1 + \Gamma' K')) = 0 && (3.3b) \\ &\nabla' c' \cdot \frac{\nabla' S}{|\nabla' S|} = (1 - c')(c' - c'_{eq}(1 + \Gamma' K')) && (3.3c) \end{aligned}\right\} S = 0$$

with appropriate initial and large $|x'|$ data. Here $\Gamma' = (k/D)\Gamma$, $c'_{eq} = c_{eq}/\rho$ and K' is the mean curvature in the $\underline{x}'$ coordinates. For notational convenience we shall now drop the primes.

A SPHERICAL SOLUTION

Spherically symmetric solutions of Equations (3.3) must necessarily have the form $c(\underline{x}, t) = c(r, t)$ and we may take $S(\underline{x}, t) = r - R(t)$ where $r = |\underline{x}|$. On the unknown surface $r = R(t)$, the mean curvature $K = 1/R(t)$. Thus Equations (3.3) for the unknowns $c(r, t)$ and $R(t)$ become

$$\frac{\partial c}{\partial t} = \frac{1}{r^2}\frac{\partial}{\partial r}\left(r^2 \frac{\partial c}{\partial r}\right) \quad r > R(t) \tag{3.4a}$$

$$\left.\begin{aligned} \dot{R}(t) &= c - c_{eq}(1 + \Gamma/R(t)) \quad &(3.4b) \\ \frac{\partial c}{\partial r} &= (1 - c)\dot{R}(t) \quad &(3.4c) \end{aligned}\right\} \quad r = R(t)$$

$$c(r, t_0) = c^0(r) \quad \text{on} \quad r > R(t_0) \tag{3.4d}$$

$$c(r, t) \to c_\infty \quad \text{as} \quad r \to \infty. \tag{3.4e}$$

To find the general solution of Equations (3.4) for arbitrary initial data is difficult. Instead, motivated by the classic Stefan problem ([6], [10]), we make an ansatz concerning the growth rate of $R(t)$ which necessitates a specific choice of initial data but permits the analytical calculation of the resulting solution. In particular, we choose

$$R(t) = \alpha\sqrt{t} \tag{3.5}$$

where α is to be determined. Substituting (3.5) into the boundary condition (3.4b) one obtains

$$\frac{\alpha}{2\sqrt{t}} = c(\alpha\sqrt{t}, t) - c_{eq} - \frac{c_{eq}\Gamma}{\alpha\sqrt{t}} \tag{3.6}$$

which suggests that we must find solutions of the radial heat Equation (3.4a) which on the boundary $R(t) = \alpha\sqrt{t}$ have specific t behavior; i.e.,

$$c(\alpha\sqrt{t}, t) = c_{eq} + \frac{1}{\alpha}\left(c_{eq}\Gamma + \frac{\alpha^2}{2}\right)t^{-1/2}. \tag{3.7}$$

The form of the relevant solutions, motivated by the similarity solutions which arise in the standard Stefan problem ([10, p. 23], for example), and the classical treatments of the heat equation [2, section 2.1, pp. 50-52], can be summarized in [3, Prop. 3.1].

PROPOSITION 1. For integral p, similarity solutions of the radial heat equation which behave like $t^{p/2}$ on $R = \alpha\sqrt{t}$ are of the form

$$c_p(r, t) = t^{p/2}(A\xi^{-1}\eta_{p+1}(\xi) + B\xi^{-1}\tilde{\eta}_{p+1}(\xi)), \; \xi = r/2\sqrt{t} \tag{3.8}$$

where η_q and $\tilde{\eta}_q$ are linearly independent solutions of

$$\eta'' + 2\xi\eta' - 2q\eta = 0. \tag{3.9}$$

Specifically,

$$\eta_q(\xi) = \begin{cases} i^q\operatorname{erfc}(\xi) = \overbrace{\int_\xi^\infty \cdots \int_\xi^\infty}^{q \text{ times}} \operatorname{erfc}(\tau)d\tau, & q > 0 \\ \operatorname{erfc}(\xi) = \int_\xi^\infty e^{-\tau^2} d\tau, & q = 0 \\ \dfrac{d^{-q}}{d\xi^{-q}} \operatorname{erfc}(\xi), & q < 0. \end{cases} \tag{3.10a}$$

while

$$\tilde{\eta}_q(\xi) = \begin{cases} \eta_q(-\xi), & q \geq 0 \\ \eta_q(\xi) \int_0^\xi \eta_q(\tau)^{-2} e^{-\tau^2} d\tau, & q < 0. \end{cases} \tag{3.10b}$$

Equation (3.7) forces us to take for $c(r, t)$ a linear combination of the solutions (3.8) with $p = 0$ and -1. Substituting this into Equation (3.4) with (3.5) and equating powers of t we obtain the following result [3, Theorem 3.3].

<u>THEOREM 2.</u> There exists a solution of the spherical equations (3.4) of the form

$$R(t) = \alpha\sqrt{t} \tag{3.11a}$$

$$c(r, t) = \frac{2\sqrt{t}}{r}\left[A\eta_1\left(\frac{r}{2\sqrt{t}}\right) + B\tilde{\eta}_1\left(\frac{r}{2\sqrt{t}}\right)\right] + \frac{2}{r}\left[C\eta_0\left(\frac{r}{2\sqrt{t}}\right) + D\tilde{\eta}_0\left(\frac{r}{2\sqrt{t}}\right)\right] \tag{3.11b}$$

where A, B, C, D are specifically calculated functions of α, c_{eq}, c_∞, Γ. The functions η_0, $\tilde{\eta}_0$, η_1 and $\tilde{\eta}_1$ are defined in Equations (3.10) and α must satisfy

$$\frac{\alpha^3}{4} e^{\alpha^2/4} \int_{\alpha/2}^\infty \tau^{-2} e^{-\tau^2} d\tau = \frac{c_\infty - c_{eq}}{1 - c_{eq}}. \tag{3.11c}$$

Using continued fraction expansions [3, Appendix 1] one can show that the left side of Equation (3.11c) is monotonically increasing from 0 to 1 on the α-interval

$[0, \infty)$. Thus there is one and only one solution of Equation (3.11c) for each set of parameters of the form $c_{eq} < c_\infty < 1$. A straightforward but lengthy calculation [3, Theorem 3.4] establishes that $c(r, t) > c_{eq}$ for $r > \alpha\sqrt{t}$ thus giving the required positivity (since it is a concentration) of the complicated expression (3.11b). Of course, solution (3.11) is a very special solution (as is the case for the similarity solution of the classic Stefan problem), but it is also a very interesting one inasmuch as it is (linearly) stable among other spherical solutions [4] and possesses the fundamental characteristic which we wanted to discuss; namely, it spontaneously loses its stability to less symmetric modes.

LINEAR STABILITY OF THE SPHERICAL SOLUTION WITH RESPECT TO BUMPS

To obtain the equations of motion for small deviations from spherically symmetric evolutions one takes

$$c_\varepsilon(r, \theta, \phi, t) = c_s(r, t) + \varepsilon c(r, \theta, \phi, t) + O(\varepsilon^2) \qquad (3.12a)$$

$$\begin{aligned} S_\varepsilon(r, \theta, \phi, t) &= S_s(r, t) + \varepsilon S(r, \theta, \phi, t) + O(\varepsilon^2) \\ &= r - R_s(t) - \varepsilon R(\theta, \phi, t) + O(\varepsilon^2) \end{aligned} \qquad (3.12b)$$

where c_s, R_s are any spherical solutions (e.g., those given in (3.11)). The equations that the first order nonsymmetric terms $c(r, \theta, \phi, t)$ and $S(r, \theta, \phi, t) = -R(\theta, \phi, t)$ must satisfy are obtained by demanding that c_ε and S_ε satisfy Equations (3.3) to first order in ε; i.e., substituting c_ε and S_ε into Equations (3.3), differentiating with respect to ε and setting $\varepsilon = 0$. Carrying out these lengthy calculations for the specific spherical solution (3.11) of the last section, one obtains [3, Corollary 4.2].

THEOREM 3. The linearization of Equations (3.3) about the spherical solution described in Theorem 2 is given by

$$\frac{\partial c}{\partial t} = \Delta c \quad \text{for} \quad r > \alpha\sqrt{t}, \quad t > t_0. \qquad (3.13a)$$

$$\left.\begin{aligned} \frac{\partial R}{\partial t} &= \left\{\frac{\alpha}{2}(1 - c_{eq})t^{-1/2} + \left[\frac{c_{eq}\Gamma}{\alpha^2} - \frac{1}{2}\left(c_{eq}\Gamma + \frac{\alpha^2}{2}\right)\right]t^{-1}\right\}R \\ &\quad + \frac{c_{eq}\Gamma}{2}\Delta R + c && (3.13b) \\ \frac{\partial c}{\partial r} &= \left\{(1 - c_{eq}) - \frac{1}{\alpha}\left(c_{eq}\Gamma + \frac{\alpha^2}{2}\right)t^{-1/2}\right\}\frac{\partial R}{\partial t} \\ &\quad + \left\{(1 - c_{eq})t^{-1} - \frac{1}{2\alpha}\left(c_{eq}\Gamma + \frac{\alpha^2}{2}\right)\right. \\ &\quad \left.\cdot\, t^{-3/2}\right\}\left(R - \frac{\alpha}{2}t^{-1/2}c\right) && (3.13c) \end{aligned}\right\} \text{ on } r = \alpha\sqrt{t},\ t > t_0$$

$$c(r,\ \theta,\ \phi,\ t_0)\ \text{given on}\ r > \alpha\sqrt{t_0} \tag{3.13d}$$

$$c(r,\ \theta,\ \phi,\ t) \to 0 \quad \text{as} \quad r \to \infty, \quad t > t_0. \tag{3.13e}$$

In order to examine the evolution of perturbations to the spherical solution associated with bumps possessing the lesser symmetry of the spherical harmonics $Y_{\ell m}(\theta, \phi)$, $|m| \leq \ell$ we take

$$R_{\ell m}(\theta,\ \phi,\ t) = \delta_\ell(t) Y_{\ell m}(\theta,\ \phi), \tag{3.14a}$$

$$c_{\ell m}(r,\ \theta,\ \phi,\ t) = c_\ell(r,\ t)\ Y_{\ell m}(\theta,\ \phi). \tag{3.14b}$$

COROLLARY 4. The solutions of the linearized Equations (3.13) which are of the form (3.14) must satisfy

$$\frac{\partial c_\ell}{\partial t} = \frac{1}{r^2}\frac{\partial}{\partial r}\left(r^2\frac{\partial c_\ell}{\partial r}\right) - \frac{\ell(\ell+1)}{r^2}c_\ell, \quad r > \alpha\sqrt{t}, \quad t > t_0 \tag{3.15a}$$

$$\dot{\delta}_\ell(t) = c_\ell(r,\ t) + \left\{\frac{\alpha}{2}(1 - c_{eq})t^{-1/2} - \frac{1}{2}\left[c_{eq}\Gamma + \frac{\alpha^2}{2} + \frac{c_{eq}\Gamma}{\alpha^2}(\ell+2)(\ell-1)\right]t^{-1}\right\}\delta_\ell(t) \tag{3.15b}$$

on $r = \alpha\sqrt{t}$, $t > t_0$

$$\frac{\partial c_\ell}{\partial r} + \left\{-(1 - c_{eq}) + \left[\frac{1}{\alpha}\left(c_{eq}\Gamma + \frac{\alpha^2}{2}\right) + \frac{\alpha}{2}\right]t^{-1/2}\right\}c_\ell$$
$$= \left\{\frac{\alpha}{2}(1 - c_{eq})^2 t^{-1/2} - (1 - c_{eq})\left[-1 + c_{eq}\Gamma + \frac{\alpha^2}{2} + \frac{c_{eq}\Gamma}{2\alpha^2}(\ell+2)(\ell-1)\right]t^{-1} + \frac{1}{2\alpha}\left(c_{eq}\Gamma + \frac{\alpha^2}{2}\right)\left[-1 + c_{eq}\Gamma + \frac{\alpha^2}{2} + \frac{c_{eq}\Gamma}{\alpha^2}(\ell+2)(\ell-1)\right]t^{-3/2}\right\}\delta(t) \tag{3.15c}$$

on $r = \alpha\sqrt{t}$, $t > t_0$

$$c_\ell(r,\ t_0)\ \text{given on}\ r > \alpha\sqrt{t_0} \tag{3.15d}$$

$$c_\ell(r,\ t) \to 0 \quad \text{as} \quad r \to \infty, \quad t > t_0. \tag{3.15e}$$

One important feature of Equations (3.15) is that the boundary is known and in particular, is the rim of the growing sphere. Motivated by the analysis used to obtain the spherical solution, we again seek solutions of Equations (3.15) for which $\delta_\ell(t)$ behaves like powers of t (specifically half-integer powers) and c_ℓ is a similarity solution of Equation (3.15a) which behaves likewise on $r = \alpha\sqrt{t}$ so that these powers of t can be matched in Equations (3.15b) and (3.15c). In particular, we look for solutions of Equation (3.15a) as sums of terms of the form

$$c_\ell(r,\ t) = t^{p/2}\ \zeta\left(\frac{r}{2\sqrt{t}}\right)\ , \tag{3.16}$$

which necessitates that $\zeta(\xi)$ satisfies

$$\zeta'' + 2(\xi^{-1} + \xi)\zeta' - \left(2p + \frac{\ell(\ell+1)}{\xi^2}\right)\zeta = 0. \tag{3.17}$$

In the radial case, $\ell = 0$, this is precisely Equation (3.9) with $q = p + 1$ after the change of variables $\eta = \xi\zeta$. Corresponding to Prop. 1, we may summarize the result in the present situation as follows [3, Prop. 5.1]:

PROPOSITION 5. The general solution of Equation (3.17) can be written in the form $A\zeta_{\ell,p}(\xi) + B\tilde{\zeta}_{\ell,p}(\xi)$ where the linearly independent parts are given by

$$\zeta_{\ell,p}(\xi) = e^{-\xi^2/2}\ \xi^{-3/2} W_{-\frac{p+3/2}{2},\ -\frac{\ell+1/2}{2}}(\xi^2) \tag{3.18a}$$

and

$$\tilde{\zeta}_{\ell,p}(\xi) = e^{-\xi^2/2}\ \xi^{-3/2}\ W_{\frac{p+3/2}{2},\ -\frac{\ell+1/2}{2}}(-\xi^2) \tag{3.18b}$$

where $W_{k,m}(z)$ is the Whittaker function [11, 16.12, p. 339].

Using the asymptotic behavior of W [11, pp. 340-3] one finds that

$$\zeta_{\ell,p}(\xi) \to e^{-\xi^2}\ \xi^{-(p+3)} \quad \text{as} \quad \xi \to +\infty \tag{3.19a}$$

$$\tilde{\zeta}_{\ell,p}(\xi) \to \xi^p \quad \text{as} \quad \xi \to +\infty. \tag{3.19b}$$

Combining formulas (3.18) and (3.19) one finds solutions

$$c_\ell(r,\ t) = At^{p/2}\ \zeta_{\ell,p}\left(\frac{r}{2\sqrt{t}}\right) + Bt^{p/2}\ \tilde{\zeta}_{\ell,p}\left(\frac{r}{2\sqrt{t}}\right) \tag{3.20}$$

of Equation (3.15a) which behave like $t^{p/2}$ on $r = \alpha\sqrt{t}$ and for each fixed $t > 0$, like $e^{-r^2}\ r^{-(p+3)}$ or r^p as $r \to \infty$. Thus for $p < 0$ we have obtained two solutions which satisfy the asymptotic condition (3.15e), while for $p \geq 0$ only the first is useful in this context.

To begin the systematic study of Equations (3.15) we estimate the maximum order of growth/decay of $\delta_\ell(t)$. As in Equations (3.11) if we choose

$$\delta_\ell(t) = dt^{p/2} \tag{3.21a}$$

then to match powers of t in (3.15b) and (3.15c) we are forced to take (with $\xi = r/2\sqrt{t}$)

$$\begin{aligned} c_\ell(r, t) &= t^{(p-1)/2}(A\zeta_{\ell,p-1}(\xi) + B\tilde{\zeta}_{\ell,p-1}(\xi)) \\ &\quad + t^{(p-2)/2}(C\zeta_{\ell,p-2}(\xi) + D\tilde{\zeta}_{\ell,p-2}(\xi)) \end{aligned} \tag{3.21b}$$

and then attempt to solve for the constants A-D in terms of d (and α, c_{eq}, c_∞, Γ). The $t^{(p-1)/2}$ coefficient in (3.15b) and the $t^{(p-2)/2}$ coefficient in (3.15c) give (writing $\zeta_{\ell,p-1}$ for $\zeta_{\ell,p-1}(\alpha/2)$, etc.) respectively

$$0 = (A\zeta_{\ell,p-1} + B\tilde{\zeta}_{\ell,p-1}) + \frac{\alpha}{2}(1 - c_{eq})d, \tag{3.22a}$$

$$\frac{1}{2}(A\zeta'_{\ell,p-1} + B\tilde{\zeta}'_{\ell,p-1}) = (1 - c_{eq})\left(\frac{p+2}{2}\right) - \frac{\alpha}{2}(A\zeta_{\ell,p-1} + B\tilde{\zeta}_{\ell,p-1}). \tag{3.22b}$$

This necessitates that for $\xi = \alpha/2$ (i.e., for all ξ in order to accommodate all c_{eq} and c_∞)

$$\begin{aligned} &A\zeta'_{\ell p-1}(\xi) + B\tilde{\zeta}'_{\ell,p-1}(\xi) \\ &\quad + [(p + 2)\xi^{-1} + 2\xi][A\zeta_{\ell,p-1}(\xi) + B\tilde{\zeta}_{\ell,p-1}(\xi)] = 0. \end{aligned} \tag{3.23}$$

Using differentiation and recursion formulas for W [1, 13.4.31 and 13.4.33, p. 507] by separate calculations for the functions with and without tildes one finds the quantity in Equation (3.23) to be

$$\xi^{-1}\left[\frac{A}{2}(p - \ell + 1)(\ell + p + 2)\zeta_{\ell,p-1}(\xi) - 2B\tilde{\zeta}_{\ell,p-1}(\xi)\right.$$

which is zero identically (since $\zeta_{\ell,p-1}$ and $\tilde{\zeta}_{\ell,p-1}$ are linearly independent) only when B = 0 (i.e., there is no solution with tilde possible in the top order term) and $p = \ell - 1$ or $-(\ell + 2)$.

The above thus suggests that for $\ell = 0$ the top order terms in $\delta_0(t)$ decay like $t^{-1/2}$). That this is indeed the case for all integrable initial data requires a very complicated proof of the sort which follows and depends on the completeness of the $\zeta_{0,-k}$ and $\tilde{\zeta}_{0,-k}$ on $[\alpha/2, \infty)$ [4].

This establishes the linear stability of our spherical solution within the class of spherical solutions, and a similar proof gives the marginal stability of the spherical solution with respect to Y_{10} bumps (as is physically expected since for small d_1, $d_1Y_{10}(\theta, \phi)$ is approximately a simple translation of the original spherical shape). Here we only sketch our results for $\ell \geq 2$. This is much simpler inasmuch as the proof of instability only requires displaying a solution which grows asymptotically. In addition, the procedure used to solve Equations (3.15) for $\ell \geq 2$

with special data is pieced together to solve Equations (3.15) for $\ell = 0$ and 1 for arbitrary data to obtain the stability results mentioned above.

THEOREM 6 ($\ell = 2$). The linearized Equations (3.15) have solutions of the form

$$\delta_2(t) = dt^{1/2} \tag{3.24a}$$

$$\begin{aligned} c_2(r, t) &= A\zeta_{2,0}\left(\frac{r}{2\sqrt{t}}\right) \\ &+ t^{-1/2}\left[C\zeta_{2,-1}\left(\frac{r}{2\sqrt{t}}\right) + D\tilde{\zeta}_{2,-1}\left(\frac{r}{2\sqrt{t}}\right)\right] \end{aligned} \tag{3.24b}$$

where the ζ's are defined in Prop. 5 and A, C, D are given in terms of α, c_{eq}, c_∞, Γ and d.

Sketch of Proof. Substituting the expressions (3.24) into Equation (3.15) (especially parts b and c) one obtains, in part, Equations (3.22) with $\ell = 2$ and $p = \ell - 1 = 1$. The subsequent discussion forced $B = 0$ (fortuitously, since $\zeta_{2,0}(r/(2\sqrt{t})) \not\to 0$ as $r \to \infty$) and hence

$$A = -\frac{\alpha}{2}(1 - c_{eq})(\zeta_{2,0})^{-1}d. \tag{3.25}$$

On the other hand, the $t^{-1/2}$ coefficient in (3.15b) and the t^{-1} coefficient in (3.15c) give, respectively,

$$C\zeta_{2,-1} + D\tilde{\zeta}_{2,-1} = \frac{d}{2}\left(1 + c_{eq}\Gamma + \frac{\alpha^2}{2} + \frac{4c_{eq}\Gamma}{\alpha^2}\right), \tag{3.26a}$$

$$C\zeta'_{2,-1} + D\tilde{\zeta}'_{2,-1} = -\frac{d}{2}\,\alpha\left(3 + c_{eq}\Gamma + \frac{\alpha^2}{2} + \frac{8c_{eq}\Gamma}{\alpha^2}\right) \tag{3.26b}$$

after a simplifying substitution using (3.26a). The system of Equations (3.26) clearly has a unique solution for C, D because the determinant of the coefficients is the Wronskian $W[\zeta_{2,-1}, \tilde{\zeta}_{2,-1}](\alpha/2) = 4e^{-\alpha^2/4}\alpha^{-2}$ (using the differential equation (3.17)).

For $\ell \geqq 3$, the above rationale is not immediately applicable. It would suggest taking

$$R_{\ell m}(\theta, \phi, t) = dt^{(\ell-1)/2}Y_{\ell m}(\theta, \phi) \tag{3.27a}$$

$$\begin{aligned} c_{\ell m}(r, \theta, \phi, t) = \Bigg\{ & t^{(\ell-2)/2}A\,\zeta_{\ell,\ell-2}\left(\frac{r}{2\sqrt{t}}\right) \\ & + t^{(\ell-3)/2}C\zeta_{\ell,c-3}\left(\frac{r}{2\sqrt{t}}\right)\Bigg\}Y_{\ell m}(\theta, \phi) \end{aligned} \tag{3.27b}$$

noting that $\tilde{\zeta}_{\ell,\ell-3}(r/(2\sqrt{t})) \not\to 0$ as $r \to \infty$. Corresponding to Equations (3.26) we would have two equations for the one unknown C. The important observation is that one can always add more slowly growing solutions to the above fundamental one. One then obtains putative solutions of the form

$$\delta_\ell(t) = dt^{-p/2} + \sum_{k=-(p+1)}^{\ell-3} d_k t^{(k+2)/2} \tag{3.28a}$$

$$c_\ell(r, t) = t^{-(p+2)/2}\left[C\zeta_{\ell,-(p+2)}\left(\frac{r}{2\sqrt{t}}\right) + D\tilde{\zeta}_{\ell,-(p+2)}\left(\frac{r}{2\sqrt{t}}\right)\right] + \sum_{k=-(p+1)}^{\ell-2} t^{k/2} A_{\ell,k}\zeta_{\ell,k}\left(\frac{r}{2\sqrt{t}}\right) \tag{3.28b}$$

and takes $p \geq -1$ so that $\tilde{\zeta}_{\ell,-(p+2)}(r/(2\sqrt{t}))$ is not precluded by the asymptotic condition. The details are given in [3, Theorem 5.3].

THEOREM 6 ($\ell \geq 3$). The linearized equations (3.15) have a solution of the form (3.28) for each $p \geq -1$.

CONCLUDING REMARKS

We have tried to summarize our work on a simple growth model ([3], [4]) showing that it admits a spherical similarity solution which loses stability to the $Y_{\ell m}$ growth modes for all $\ell \geq 2$. By taking $p > 1$ in (3.28) $\tilde{\zeta}_{\ell,-(p+2)}$ is integrable at infinity and the ζ's being so trivially shows how to obtain the same results with "finite mass" perturbations. For $\ell = 0$ and 1, the same idea with $p \to \infty$ is the basis of the stability proof. From the form of (3.28a), it is clear that the notion of a critical radius at which the spherical growth loses stability to the $Y_{\ell m}$ modes is more complicated than in the quasi-equilibrium theory ([8], [9], [7]). This, a mode-mode analysis of the nonlinear equations (3.3) as well as the coupling of this model to an external electric field [5] are some of the topics of our current research in this area.

REFERENCES:

1. Abramowitz, M. and Stegun, I. A., Handbook of Mathematical Functions, U.S. Government Printing Office, Washington, D.C., 1964.

2. Carslaw, H. S. and Jaeger, J. C., Conduction of Heat in Solids, Clarendon, Oxford, 1959.

3. Chadam, J. and Ortoleva, P., Morphological instability of growing amorphous solids, 54 mimeo pages, submitted for publication (1980).

4. Chadam, J. and Ortoleva, P., in preparation.

5. Chadam, J., ElBaudewi, M. and Ortoleva, P., in preparation

6. Friedman, A., Partial Differential Equations of Parabolic Type, Prentice-Hall, Englewood Cliffs, New Jersey, 1964.

7. Langer, J. S., Instabilities and pattern formation in crystal growth, Carnegie-Mellon preprint (1980).

8. Mullins, W. W. and Sekerka, R. F., Morphological stability of a particle growing by diffusion or heat flow, J. Appl. Phys. 34: 323 (1963).

9. Mullins, W. W. and Sekerka, R. F., Stability of a planar interface during solidification of a dilute binary alloy, J. Appl. Phys. 35: 444 (1964).

10. Rubenstein, L. I., The Stefan Problem, Vol. 27, Translations of Mathematical Monographs, AMS, Providence, Rhode Island, 1971.

11. Whittaker, E. T. and Watson, G. N., A Course on Modern Analysis, University Press, Cambridge, 1963.

MATHEMATICAL TREATMENT OF NERVE CONDUCTION AND CARDIAC PURKINJE FIBERS

Jane Cronin
Department of Mathematics, Rutgers University,
New Brunswick, New Jersey

INTRODUCTION

The distinguished work of Hodgkin and Huxley [7, 8, 9, 10, 11] consists of two parts: the use of novel and delicate experimental techniques for obtaining qualitative data on the electrical activity of the squid giant axon; and a description and summary of these data in terms of a system of nonlinear ordinary differential equations, i.e., a model of the electrical activity of the squid giant axon. These experimental and mathematical techniques have been adapted and extended for the study of a number of electrically active cells including the lobster giant axon (Julian, Moore, Goldman [12]), a myelinated axon (Frankenhaeuser and Huxley [5]), cardiac Purkinje fibers (McAllister, Noble and Tsien [14]) and ventricular myocardial fibers (Beeler and Reuter [2]).

Since the mathematical models derived for these various cells are all obtained by using the approach of Hodgkin and Huxley, i.e., by introducing activation and inactivation variables to describe the current flow, across the membranes, composed of various ions, it is not surprising that the models for the various cells are all systems of nonlinear ordinary differential equations and that these systems bear a marked resemblance to one another. However the physiological functions of the different cells are very different and, consequently, the mathematical problems which must be considered are also very different.

This paper has two purposes: first to describe and contrast some of the widely varying problems which arise in the study of two models: those of the squid giant axon and the cardiac Purkinje fiber; and second, to describe the status of the problems which arise in the model of the cardiac Purkinje fiber. Thus our discussion will consist largely of stating problems and only to a small extent of indicating solutions. There is, however, a valid reason for emphasizing problems rather than solutions. We are concerned here with mathematical problems which arise from the underlying physiology, not with a mathematical study of, say, the Hodgkin-Huxley equations for their own sake. Hence it is important to emphasize that the mathematical description given by the Hodgkin-Huxley equations (and the other models cited above) are empirical and tentative (see Hodgkin and Huxley [11, page 541]). Ongoing physiological studies will probably yield better models within ten or fifteen years. Work done on the formulation of the mathematical problems which arise from the underlying physiology has a good chance of being applicable to more realistic models. Whether results on the solutions of the present models will carry over to future models remains to be seen.

THE SQUID GIANT AXON

The purpose of the work of Hodgkin and Huxley was to understand how a nerve impulse is conducted along an axon, and we begin our discussion by considering the Hodgkin-Huxley equations themselves. We will write them in the form given by FitzHugh [4]. In this form, an outward flow of positive ions is a positive current (this agrees with present practice in cell electrophysiology), but the resting potential is taken as the zero for the membrane potential (in present practice in cell electrophysiology, the resting potential is about -70 mV). Let V denote the potential difference across the membrane, let m, h be the activation and inactivation variables for the flow of sodium ions and let n be the activation variable for the flow of potassium ions. We will not discuss the meanings of m, h and n except to say that m and h are used to measure the permeability of the membrane to sodium ions and n is used to measure the permeability of the membrane to potassium ions. Thus m, h, n describe indirectly the flow of sodium and potassium ions across the axon membrane. It should be emphasized, however, that the introduction and use of these activation variables by Hodgkin and Huxley was both ingenious and profound. A measure of that profoundity is the fact that these activation and inactivation variables have been used ever since, not only in the mathematical description of the squid axon but also in the descriptions of other electrically active cells such as those cited in the introduction.

In the notation just described, the Hodgkin-Huxley equations are:

$$I = C\dot{V} + I_i$$

$$I_i = \bar{g}_{Na} m^3 h (V - V_{Na}) + \bar{g}_K n^4 (V - V_K) + \bar{g}_L (V - V_L)$$

$$\dot{m} = \alpha_m(V)[1 - m] - \beta_m(V)m = (m_\infty(V) - m)/\tau_m(V)$$

$$\dot{h} = \alpha_h(V)[1 - h] - \beta_h(V)h = (h_\infty(V) - h)/\tau_h(V)$$

$$\dot{n} = \alpha_n(V)[1 - n] - \beta_n(V)n = (n_\infty(V) - n)/\tau_n(V)$$

where

V = displacement of the membrane potential from the rest value

I = total current through membrane, per unit area

I_i = ionic current through membrane, per unit area

C = capacitance per unit area of membrane

$\bar{g}_{Na}$ = sodium conductance constant = 120 mmho/cm^2 (here mho = 1/ohm, the reciprocal of the ohm and the prefix m denotes "milli")

$\bar{g}_K$ = potassium conductance constant = 36 mmho/cm^2

$\bar{g}_L$ = leakage conductance constant = 0.3 mmho/cm^2

V_{Na} = 115 mV

V_K = -12 mV

V_L = 10.5988 mV

$$\alpha_m(V) = \frac{.1(25 - V)}{e^{0.1(25-V)} - 1}$$

$$\beta_m(V) = 4e^{\frac{V}{-18}}$$

$$\alpha_h(V) = .07\ e^{\frac{V}{-20}}$$

$$\beta_h(V) = \frac{1}{e^{.1(30-V)} + 1}$$

$$\alpha_n(V) = \frac{.01(10 - V)}{e^{.1(10-V)} - 1}$$

$$\beta_n(V) = .125\ e^{\frac{V}{-80}}$$

Hodgkin and Huxley [11] described successes of their equations in predicting experimentally observed physiological phenomena, and other such successes were obtained later. For a summary of these, see FitzHugh [4, pp. 31-32]. The Hodgkin-Huxley equations also have some drawbacks in that certain of the predictions made by the equations do not agree with experimental results. For example, numerical analysis of the equations suggests the existence of periodic solutions if I is a positive value in certain ranges. But the corresponding experimental results are not oscillatory. See FitzHugh [4, page 32]. A modification of the Hodgkin-Huxley equations has been proposed by Adelman and FitzHugh [1] which gives better predictions. That is, numerical analysis of the modified equations does not suggest the existence of such periodic solutions.

Now we describe a drawback of Hodgkin-Huxley equations which cannot be remedied by modifying the differential equations. In understanding how a nerve impulse travels along an axon, the concept of threshold phenomenon plays a prominent role. If a stimulus is applied to the axon and if the magnitude of the stimulus exceeds a certain fixed value, called the threshold value, then the stimulus initiates a nerve impulse or action potential. (The axon is said then to "fire.") If the amplitude of the stimulus is less than the threshold value, no nerve impulse is produced. An important purpose of the mathematical model is to describe how this threshold phenomenon occurs, and we will discuss one aspect of this problem. It should be emphasized that similar discussions would be valid for other physiological systems for which the threshold phenomenon is important, e.g., the myelinated axon.

We point out first that the Hodgkin-Huxley equations give a good description of the threshold phenomenon. That is, a solution of the equations with an initial value (at $t = 0$) $V = V_o$, where V_o is greater than or equal to the threshold value and initial values $m = m_o$, $h = h_o$, $n = n_o$ where m_o, h_o, n_o are the values of m, h, n when the axon is at rest (i.e., $m_o = m_\infty(0)$, $h_o = h_\infty(0)$, $n_o = n_\infty(0)$) describes, as t increases above zero, the changes that take place when an action potential occurs. But if the initial values at $t = 0$ are such that V_o is less than the threshold value and $m = m_\infty(0)$, $h = h_\infty(0)$, $n = n_\infty(0)$, then the solution describes the changes that take place when no action potential occurs, i.e., when there are only "passive" changes.

Since the functions which appear in the Hodgkin-Huxley equations are continuously differentiable functions, the standard existence and uniqueness theorems for solutions of the equations are applicable. Hence a solution which satisfies the initial conditions $V(0) = V_o$, $m_o = m_\infty(0)$, $h_o = h_\infty(0)$, $n_o = n_\infty(0)$ is unique. Thus if V_o is the threshold value, the prediction given by the differential equation is unique: an action potential always occurs, i.e., the nerve always fires. But this prediction does not agree with experimental observations. In experiments, if $V(0)$ is set at the threshold value, the axon does not always respond by firing. Firing occurs only for a certain proportion of the number of times that stimulus occurs. The reason for the varying responses of the real axon to threshold stimulus is thought to be the random fluctuations or "noise" in the resting potential. This suggests that the Hodgkin-Huxley equations would give a more realistic description of the behavior of the axon if they contained a random or stochastic term. Also the study of the Hodgkin-Huxley equations modified by the addition of such a stochastic term would make possible a theoretical connection between probabilistic firing of axons and electrical noise generated across the nerve membrane. It should be emphasized at this point that the addition of a stochastic term is not just for the purpose of constructing a novel mathematical exercise. Random firing of axons exposed to near-threshold stimuli has been studied by physiologists for many years. It is a serious subject in physiology and is deserving of a quantitative description.

On the other hand if the Hodgkin-Huxley equations are modified by the addition of a stochastic term, then we pass from the realm of ordinary differential equations for which there is an extensive general theory to the subject of stochastic differential equations which is a newer and less fully developed branch of analysis.

A study in this direction has been carried out by Lecar and Nossal [13]. They used the 2-dimensional simplification of the Hodgkin-Huxley equations proposed by FitzHugh [4] and described the dynamics of a noisy nerve with a 2-dimensional brownian motion. Their analysis was carried out by linearizing the FitzHugh equations in the neighborhood of the unstable equilibrium point (a saddle point). Lecar and Nossal used their approach to study aspects of excitability fluctuation which have been investigated experimentally and their theoretical results agree with experimental results. Moreover, addition of the stochastic term improves the FitzHugh model in a somewhat unexpected way. In the FitzHugh model, the two dependent variables are m and V, and the saddle point is used to model the threshold. Inspection of the simple case of a saddle-point in Figure 4.1 shows how the

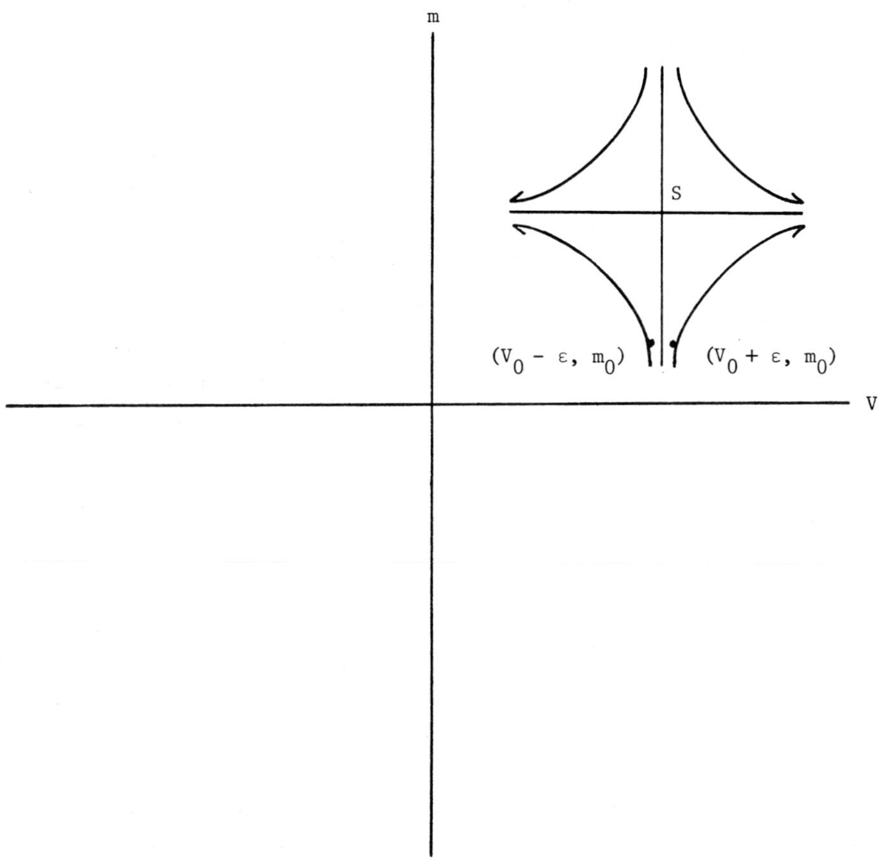

Fig. 4.1.

threshold can be described. If the solution passes through the point $(V_o - \varepsilon, m_o)$ then the solution veers to the left as indicated in Figure 4.1. But if a solution passes through the point $(V_o + \varepsilon, m_o)$ then the solution veers to the right, and these different behaviors of the solution curves occur no matter how small the positive number ε is. Thus the number V_o plays the role of a threshold value. (For a more detailed and realistic description of the saddle-point threshold model, see FigzHugh [4, pp. 36-37].) However as FitzHugh [4, pp. 37-38] points out, there is a drawback to the saddle-point threshold model. If we consider a solution which passes through the point $(V_o + \varepsilon, m_o)$ where ε is very small then the solution comes very close to the saddle-point S itself. Since S is an equilibrium point, motion along a solution curve very near S takes place very slowly. If ε is sufficiently small, then an arbitrarily large value of t must be attained before the solution curve begins to veer away from the saddle point. This means, in physiological terms, that an arbitrarily long period of time may elapse between application of stimulus and the firing of the nerve. The time interval between application of stimulus and firing of the nerve is called the latency, and we may summarize the preceding remarks by saying that the saddle-point threshold model has unbounded latency. But experimental results indicate that latency is bounded. Thus the unbounded latency of the saddle-point threshold model is one of its drawbacks. However Lecar and Nossal show that the addition of the stochastic term makes the model more realistic because there are very few long-latency trajectories.

Since our primary purpose is to point out problems and directions of study rather than to summarize work that has been done, we will not give a detailed description of the results of Lecar and Nossal. We restrict ourselves to pointing out that a realistic mathematical treatment of the threshold phenomenon seems to require stochastic considerations.

THE CARDIAC PURKINJE FIBER

Extensive experimental work has shown that the electrical activity of the cardiac Purkinje fiber is, in certain respects, similar to that of the squid axon, and a quantitative description or mathematical model of this electrical activity consists of a system of nonlinear ordinary differential equations which are analogous to the Hodgkin-Huxley equations. As we shall see later, a major difference between the two systems arises from the fact that the number of ionic currents across the membrane in the cardiac Purkinje fiber is much larger than the number of ionic currents across the membrane in the squid axon. But our first step is to describe briefly the functions of the cardiac Purkinje fiber. The initiation of the electrical activity which governs the heartbeat is carried on by a pacemaker mechanism which is located at a position in the heart called the sino-atrial node (SA node). (The SA node is not the only region of the heart which possesses a pacemaker mechanism. But its rate of beating is higher than that of any other pacemaker region, and so it sets the pace of the heart as a whole.) The electrical impulses generated by the pacemaker mechanism are conducted to the various parts of the heart, first to the atria, then through a small strip of tissue called the atrio-ventricular (AV) node and finally to the ventricles. The ventricles are the two large chambers of the heart into which blood is brought from the two smaller thin-walled chambers called the atria and from which the blood is pumped to lung circulation and systemic circulation. The electrical impulse is conducted comparatively slowly through the AV node. Then it is conducted more rapidly to all of the walls of the ventricles. The electrical impulse causes the muscles in the ventricular walls to contract thus producing the heartbeat and pumping the blood into the circulatory system. Consequently it is important that the muscles in the ventricular walls contract almost simultaneously so that the blood is forced into the arteries at high pressure. Thus, in turn, it is important that the electrical impulse be conducted very rapidly from the AV node to the ventricular walls. The electrical impulses are conducted from the AV node to the ventricular walls by the cardiac Purkinje fibers, and their large size insures that the conduction is rapid.

Conduction of electrical impulses that originate in the pacemaker region is the primary function of the cardiac Purkinje fibers, but they have also a secondary function. The Purkinje fibers fire spontaneously at a regular rate. This regular rate is slower than the rate of firing of the pacemaker mechanism and consequently under normal conditions (i.e., when the Purkinje fiber conducts an electrical impulse which originates from the pacemaker mechanism) the Purkinje fiber does not fire spontaneously. However if the electrical impulses from the pacemaker mechanism do not reach the Purkinje fibers (this happens, for example, in a clinical condition called AV block) then the spontaneous firing of the Purkinje fibers plays a major role. The Purkinje fibers then act as a pacemaker, and the heart beats at the rate of the spontaneous firing of the Purkinje fibers. Thus the secondary function of the Purkinje fibers is to act as a "back-up" pacemaker.

The experimental techniques (the voltage-clamp methods) developed by Hodgkin and Huxley for the study of the squid axon have been extended and adapted for the study of Purkinje fibers. Developing voltage clamp methods for the study of Purkinje fibers was by itself difficult experimental work, and a mathematical model based on the experimental work was not obtained until 1975 when the work of McAllister, Noble and Tsien [14] was published.

The equations derived by McAllister, Noble and Tsien to model electrical activity in the Purkinje fiber are similar to the Hodgkin-Huxley equations in that there is an equation which describes the rate of change of potential difference across the surface of the Purkinje fiber, and there is a set of equations describing the rates of change of activation and inactivation variables associated with the various ionic currents. The equations, which will be referred to hereafter as the MNT equations, are as follows:

$$\begin{aligned} 0 = C\frac{dV}{dt} &+ [\bar{g}_{Na}(V - V_{Na})m^3h] + \bar{g}_{Si}(V - V_{Si})df \\ &+ \bar{g}'_{Si}(V - V_{Si})d' + [\bar{i}_{K_2}(V)]s + [\bar{i}_{x_1}(V)]x_1 \\ &+ [\bar{i}_{x_2}(V)]x_2 + qr(\bar{g}_{qr}\ [V - V_{C\ell}] \\ &+ i_{K_1}(V) + i_{Na,b}(V) + i_{C\ell,b}(V) \end{aligned}$$

$$\frac{dy}{dt} = [\alpha_y(V)](1 - y) - \beta_y(V)]y$$

$$(y = m,\ d,\ s,\ x_1,\ x_2,\ q,\ h,\ f,\ r)$$

where

C is membrane capacitance, a positive constant

V is the potential difference across the surface of the Purkinje fiber

$[\bar{g}_{Na}(V - V_{Na})m^3h] = i_{Na}$ = inward current of Na ions where $\bar{g}_{Na}$, V_{Na} are constants and m, h are activation and inactivation variables respectively

$[\bar{g}_{Si}(V - V_{Si})df] + \bar{g}'_{Si}(V - V_{Si})d' = i_{Si}$ = inward current consisting partly of Ca ions, where $\bar{g}_{Si}$, V_{Si}, $\bar{g}_{Si}$ are constants and d, f are activation and inactivation variables respectively and d' is a function of V

$[\bar{i}_{K_2}(V)]s = i_{K_2}$ = outward K current, where $\bar{i}_{K_2}(V)$ is a function of V, s is an activation variable

$[\bar{i}_{x_1}(V)]x_1 = i_{x_1}$ = another outward K current where x_1 is activation variable

$[\bar{i}_{x_2}(V)]x_2 = i_{x_2}$ = third outward K current where x_2 is activation variable

$[qr(\bar{g}_{qr})(V - V_{C\ell})] = i_{qr}$ = outward chloride current where $\bar{g}_{qr}$, $V_{C\ell}$ are constants, and q, r are activation and inactivation variables respectively

$i_{K_1}(V)$ = outward K current depending only on V

$i_{Na,b}(V)$ = inward Na current depending only on V

$i_{C\ell,b}(V)$ = outward chloride current depending only on V

$\alpha_y(V)$, $\beta_y(V)$ are functions of V for y = m, d, s, x_1, x_2, q, h, f, r.

Since we do no calculations with these equations, we will not specify the various functions of V or give the values of the various constants. This information is, of course, given in [14]. We merely point out that all the functions are continuously differentiable (so that the usual existence and uniqueness theorems for the solutions of ordinary differential equations are applicable). Note also that this is a 10-dimensional system of differential equations.

The formation of an action potential in the Purkinje fiber is roughly similar to the formation of an action potential in the squid axon. When the membrane potential is at a suitable level (equal to or above the threshold level) the current i_{Na} begins to increase. Then i_{Si} increases more slowly, i_{qr} develops and i_{Na} decreases. Then the membrane potential stays on a plateau. Finally i_{X_1} starts, followed by i_{X_2} and i_{K_2}. Up to this stage, the pattern is roughly similar to that which occurs with the squid axon. But the next events which occur are qualitatively different. The current i_{K_2} decreases and an inward background current (a current which depends only on V and consists primarily of an inward flow of sodium ions) becomes influential. This inward background current raises the membrane potential above the threshold value and the entire sequence of events, starting with the increase of the current i_{Na}, is repeated. Thus we have a rough description of the regular spontaneous firing of the Purkinje fiber, i.e., the secondary or pacemaker function of the Purkinje fibers. Since the decrease of the current i_{K_2} permits the flow of background current to raise the membrane potential above the threshold value and hence to initiate another action potential, the current i_{K_2} is called the pacemaker current.

It should be noticed that, unlike the electrical activity in the squid axon, the threshold plays a much less important role in the functioning of the Purkinje fiber. In the Purkinje fiber, the background current raises the membrane potential above the threshold value and continues to increase the membrane potential until firing occurs, i.e., until the current i_{Na} begins to increase. This suggests that it is not needful to include the kind of stochastic term that was added to the Hodgkin-Huxley equations. That is, to obtain a quantitative description of the functions of the Purkinje fiber, it is sufficient to use a system of ordinary differential equations, the MNT equations.

Now we turn to mathematical analysis of the MNT equations. The numerical analysis of the MNT equations carried out in the original paper [14] suggests the existence of an oscillatory solution. Indeed, it seems reasonable that the regular spontaneous firing of the Purkinje fibers would be described by a periodic solution of the MNT equations. We would expect, moreover, that this periodic solution would be asymptotically stable with a fairly large region of stability. ("Asymptotically stable" here means phase asymptotically stable. A solution $x(t)$ of an autonomous system $\dot{x} = f(x)$ is phase asymptotically stable iff $\varepsilon > 0$ implies there exists $\delta > 0$ such that if $y(t)$ is a solution and if there exists $t_1 > 0$, $t_2 > 0$ such that

$$|x(t_1) - y(t_2)| < \delta$$

then for all $t > 0$,

$$|x(t_1 + t) - y(t_2 + t)| < \varepsilon$$

and there exists t_3 such that

$$\lim_{t\to\infty} |x(t + t_3) - y(t)| = 0 .)$$

Thus, the first mathematical problem to be studied for the MNT equations is a very classical question: to establish the existence of an asymptotically stable periodic solution. Let us assume that this problem has been solved. Next we consider the mathematical description of the primary function of the Purkinje fibers, i.e., the transmission of regular electrical impulses which are initiated in the pacemaker region (at the SA node) and which occur with a higher frequency than the frequency of the spontaneous firings of the Purkinje fiber. In order to obtain such a mathematical description, we must add to the MNT equations a term describing the periodic electrical impulses which originate in the pacemaker. If we write the MNT equations in vector form as

$$\dot{x} = P(x) \tag{1}$$

and if T is the period of the asymptotically stable periodic solution of the MNT equations the existence of which is assumed to have been established, then the pacemaker impulses can be described by adding to (1) a term of the form $F(t, \varepsilon)$ where $F(t, \varepsilon)$ has period $T - \varepsilon$ and $F(t, 0) \equiv 0$ and where ε is a positive number. Then the second mathematical problem to be studied for the MNT is also a very classical question: to determine if the equation

$$\dot{x} = P(x) + F(t, \varepsilon) \tag{2}$$

has a solution of period $T - \varepsilon$, i.e., to determine if entrainment of frequency takes place.

We have shown that the modeling of the primary and secondary functions of the Purkinje fibers become classical mathematical problems concerning systems of nonlinear ordinary differential equations. Now we discuss briefly the possibility of solving these problems. The first problem, to establish the existence of an asymptotically stable periodic solution, is a very old problem which is mostly unsolved. Just to establish the existence of a periodic solution is a difficult problem. The Poincaré-Bendixson Theorem solves the problem in the case $n = 2$, but for $n > 2$, the known results are fragmentary. Most hold only for some special case. See, for example, Hastings, Tyson, and Webster [6]. Even if the existence of a periodic solution is established the problem of showing that the periodic solution is asymptotically stable is by itself difficult. A big obstruction to answering the stability question is that very often the existence of the periodic solution is established by a qualitative argument, e.g., a fixed point theorem, and consequently very little information concerning the periodic solution is obtained, especially information which would indicate whether the solution is stable. For example, there may be no way to obtain estimates of the characteristic multipliers associated with the periodic solution. There is a second approach to the problem of finding an asymptotically stable periodic solution based on a theorem of George Sell. Sell's Theorem says roughly that a bounded asymptotically stable solution approaches (has as its Ω-limit set) an asymptotically stable periodic solution (which may be an equilibrium point). Thus if the existence of an asymptotically stable solution can be established and if there are no asymptotically stable equilibrium points, then it follows that there exists an asymptotically stable periodic solution which is not an equilibrium point. For discussions of this, see Cronin [3].

The second mathematical problem, to determine if equation (2) has a solution of period $T - \varepsilon$, is also an old problem and there are standard methods for establishing the existence of such periodic solutions and determining their stability properties. The basis for the methods is a technique introduced and developed by Poincaré. There are many studies of this method and its applications. For general discussions of the method, see Malkin [15] and Cronin [3]. However in order to apply the method, it is necessary first to establish the existence of an asymptotically stable periodic solution of the autonomous system, i.e, the MNT model. Thus it is imperative to solve the first mathematical problem that was formulated.

We add two remarks to this brief description of the mathematical model of the Purkinje fiber. First, we emphasize explicitly a point that has already been made. Although there is a formal similarity between the Hodgkin-Huxley equations and the MNT equations, the mathematical analysis that must be undertaken in the study of the two models are very different. A mathematical treatment of threshold phenomena seems to require stochastic considerations, but mathematical treatment of the functions of the Purkinje fiber seems to require only the solution of very classical problems in nonlinear ordinary differential equations. Finally, because the various ionic currents across the surface of the Purkinje fiber are activated at different rates, the MNT equations can be approximated by a singularly perturbed system of nonlinear ordinary differential equations. (This kind of observation has often been made about mathematical models of electrically active cells. See, for example FitzHugh [4, pages 15-16].) From this viewpoint, the two problems that have been posed for the MNT equations are close analogs of the problems posed by van der Pol and van der Mark [16] in 1928 in their discussion of relaxation oscillations.

REFERENCES

1. Adelman, W. J. and FitzHugh, R., Solutions of the H - H equation modified for potassium accumulation in a periaxonal space, Federation of American Societies for Experimental Biology Proceedings 34: 1322 (1975).

2. Beeler, G. W. and Reuter, H., Reconstruction of the action potential of ventricular myocardial fibers, J. Physiology 268: 177 (1977).

3. Cronin, J., Differential Equations: Introduction and Qualitative Theory, Marcel Dekker, New York, 1980.

4. FitzHugh, R., Mathematical models of excitation and propagation in nerve, Chapter 1, Biological Engineering, H. B. Schwan, editor, McGraw-Hill, New York, 1969.

5. Frankenhaeuser, B. and Huxley, A. F., The action potential in the myelinated nerve fibre of Xenopus laevis as computed on the basis of voltage clamp data, J. Physiology 171: 302 (1964).

6. Hastings, S. P., Tyson, J. J. and Webster, D., Existence of periodic solutions for negative feedback cellular control systems, J. Differential Eqns. 25: 39 (1977).

7. Hodgkin, A. L., Huxley, A. F. and Katz, B., Measurement of current-voltage regulations in the membrane of the giant axon of Loligo, J. Physiology 116: 424 (1952).

8. Hodgkin, A. L. and Huxley, A. F., Currents carried by sodium and potassium ions through the membrane of the giant axon of Loligo, J. Physiology 116: 449 (1952).

9. Hodgkin, A. L. and Huxley, A. F., The components of membrane conductance in the giant axon of Loligo, J. Physiology 116: 473 (1952).

10. Hodgkin, A. L. and Huxley, A. F., The dual effect of membrane potential on sodium conductance in the giant axon of Loligo, J. Physiology 116: 497 (1952).

11. Hodgkin, A. L. and Huxley, A. F., A quantitative description of membrane current and its application to conduction and excitation in nerve, J. Physiology 117: 500 (1952).

12. Julian, F. J., Moore, J. W. and Goldman, D. E., Current voltage relations in the lobster giant axon membrane under voltage-clamp conditions, J. Gen. Physiol. 45: 1217 (1962).

13. Lecar, H. and Nossal, R., Theory of threshold fluctuations in nerves. I. Relationships between electrical noise and fluctuations in axon firing, Biophysical J. 11: 1048 (1971).

14. McAllister, R. E., Noble, D. and Tsien, R. W., Reconstruction of the electrical activity of cardiac Purkinje fibers, J. Physiology 251: 1 (1975).

15. Malkin, I. G., Some Problems in the Theory of Nonlinear Oscillations, translated by United States Atomic Energy Commission Technical Information Service, 1959.

16. van der Pol, B. and van der Mark, J., The heartbeat considered as a relaxation oscillation, and an electrical model of the heart, Philos. Mag. 7th Ser. 6: 763 (1928).

A NUMERICAL METHOD FOR PHYSIOLOGICAL FLOWS

S. K. Dey
Department of Mathematics, Eastern Illinois University,
Charleston, Illinois

ABSTRACT

Several models of mathematical physiology are basically problems of fluid dynamics represented by a set of nonlinear parabolic partial differential equations whose analytical solutions are yet to be known. These equations are generally solved by explicit finite difference methods which have, in general, poor stability properties. Although implicit finite difference methods have better stability properties actual computations involve complicated algorithms requiring large memory storage in a computer. In this article a simple numerical method has been described which could efficiently solve implicit finite difference analogs of nonlinear partial differential equations. This method is known as PIS (Perturbed Iterative Scheme) and is obtained by adding a perturbation parameter to nonlinear Gauss-Siedel iterations. Preliminary tests indicate that this method, which requires no Jacobians to be computed and no linearization at any stage, could efficiently solve certain nonlinear parabolic partial differential equations representing viscous fluid flow. The algorithm is simple and computer programming is rather easy.

INTRODUCTION

Laminar, viscous fluid flow is described by nonlinear second order partial differential equations, known as Navier-Stokes' equations, whose analytical solutions subject to general initial-boundary conditions, are yet to be known. These equations are solved now by computational methods and with the development of high speed computers with large memory storage capabilities better computational schemes are being employed to obtain numerical solution. These equations are sometimes used in physiology to study various characteristics of flow properties in the human body. One such very important flow phenomenon is the pulsatile flow of blood. But blood is no ordinary fluid, and its motion through our veins and arteries represent one of the most complicated flow phenomena of fluid dynamics. Let us first briefly discuss certain unique features of blood and its motion.

Blood is a suspension of cells in a colorless liquid known as blood plasma. These cells consist of (i) red cells (erythocytes) which form the majority of the blood cells, (ii) the white cells (leukocytes) and (iii) platelets (thrombocytes). Plasma contains 91% water, about 6.5% protein, about 1.4% substances like glucose, urea transported by the blood stream and about 0.6% inorganic salt. When blood is exposed to air, certain biochemical reactions occur in the formation of fibrin and blood coagulation takes place. To obtain some reliable values of blood viscosity a large number of experiments were conducted on blood flow varying hematrocits, anticoagulants, temperatures, etc. It has been found [19] that for low rates of strain blood behaves like a non-Newtonian fluid (Figure 5.1a). To conduct some stress-strain relation for blood, Charm et al. [4] fitted Casson's equation and showed

$$\sqrt{\tau} = 0.166\sqrt{\dot{\gamma}} + 0.33(\text{dynes/cm}^2)^{1/2}$$

where τ = the shear stress, $\dot{\gamma}$ = shear rate of strain. The formula is assumed to be valid for $\dot{\gamma} = 0$ to 10^5 sec.$^{-1}$.

This unusual fluid, blood, displays an unusually complicated motion. The motion is pulsatile and through distensible passageways. In the Figure 5.1b principle elements of the arterial system are shown. It shows extremely intricate and complex branched networks of tubes. Through these tapered blood vessels flow is generally laminar with a peak Reynolds number about 10^4. However, due to any occlusion or stenosis in the circulatory system this laminar flow could become turbulent. Other complexities of the flow occur when "entry-region-flow" is considered or secondary-flow effects caused by centrifugal action in curved tubes are taken into consideration. In [17] Lighthill discussed most of these problems in hemodynamics in great details.

Mathematical description of the incredibly complicated motion of the unusually complex fluid, blood, is a formidable task. Still inquiries and interests to reveal the properties of the in vivo flow of blood led to some mathematical study of this flow using simplified physiological models. But most of these models consist of nonlinear, second order, partial differential equations whose analytical solutions are yet to be known. Hence numerical techniques are used for some approximate solution. Most of these numerical schemes pose serious computational problems related to stability and convergence. We will now consider two models on blood flow and study the nature of the computational difficulties involved with their numerical solutions.

Model I. Blood Flow in the Entrance Region of a Tube

Interest in this area has grown because of the development of capillary membrane oxygenators which separate blood and gas phase with semipermeable membranes.

The equations representing conservation of mass and momentum for incompressible, laminar, steady, axisymmetric boundary layer flow are given by:

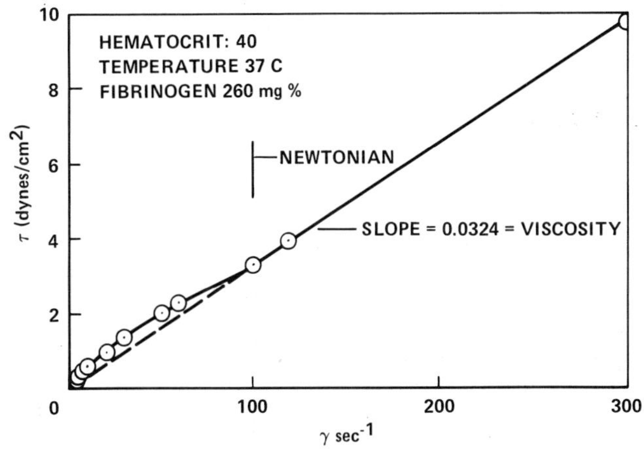

SHEAR STRESS τ IS SHOWN AS A FUNCTION OF STRAIN RATE γ FOR NORMAL BLOOD FROM MERRILL [19]

Fig. 5.1a.

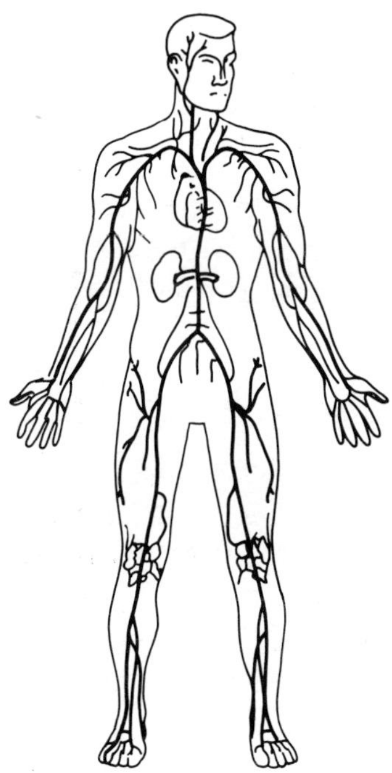

SCHEMATIC OF THE ARTERIAL SYSTEM

Fig. 5.1b.

$$\frac{\partial}{\partial x}(ru) + \frac{\partial}{\partial r}(rv) = 0 \quad (5.1)$$

$$u\frac{\partial u}{\partial x} + v\frac{\partial u}{\partial r} = -\frac{1}{\rho}\frac{dp}{dx} + \frac{1}{\rho r}\frac{\partial}{\partial r}(\tau r) \quad (5.2)$$

where

u = velocity component along the axial (x) direction

v = velocity component along the radial (r) direction

ρ = density

p = pressure

τ = shear stress given by Casson's formula

$$\tau = (\sqrt{\tau_y} + K_c\sqrt{\dot{\gamma}})^2 \quad (5.3)$$

τ_y = yield stress, $\dot{\gamma}$ = shear rate of strain

K_c = a constant.

The boundary conditions in the entrance of the tube are:

$$u = v = 0 \quad \text{at} \quad r = r_0 \qquad \partial u/\partial r = 0 \quad \text{at} \quad r = 0$$

and $u = \bar{u}$ at $x = 0$. (r_0 = radius of the tube and $\bar{u}$ is the average velocity.) This model was solved by Shah and Soto [29] using an explicit finite difference scheme discussed by Patankar and Spalding [24].

Here motion is assumed to be steady and only validation of the numerical results was obtained when the flow is considered to be fully developed. (From the physiological point of view fully-developed and steady flow of blood is a rather vague concept!) Another validation of numerical solution was done against some available experimental data when the motion is Newtonnian. All these validations apparently show success of the computational procedure.

Validation of most of these assumptions is outside the scope of numerical analysis. However, one may see that the assumption of a steady flow made it easier for the authors to apply an explicit finite difference scheme for numerical solution. This led to the consideration of a fully-developed flow and its validation against available results. If flow is considered to be unsteady, explicit schemes, which require small increment of time for numerical stability will require a large amount of computational time before steady-state solution may be obtained. Although, implicit finite difference methods generally display better stability properties even with large time steps, they have in general complicated algorithms and require large computer-memory-storage and computational time. Thus there is a need for an implicit finite difference method with a built-in acceleration parameter for faster convergence rate and requiring less computer-memory storage.

Computational difficulties possibly led some researchers like Lew [16] even to linearize the equation of motion (5.2) to obtain some analytical solution.

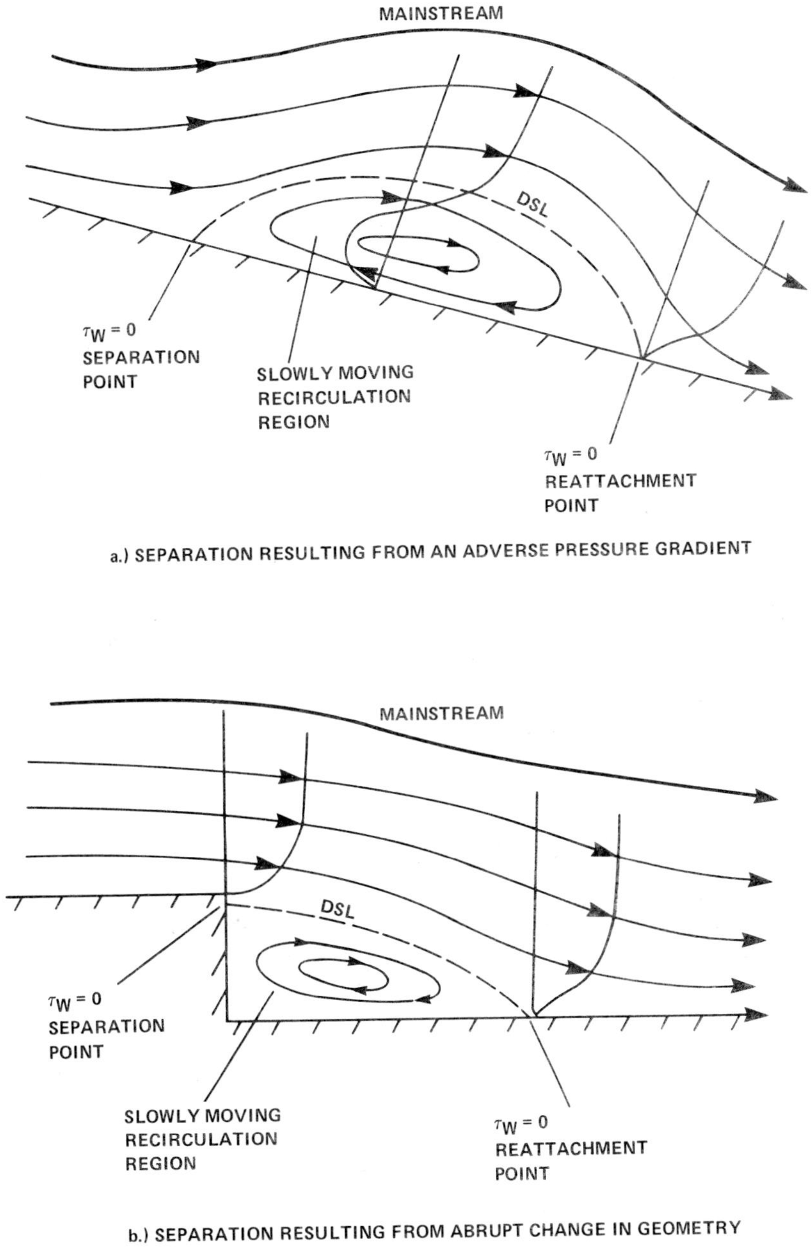

Fig. 5.2. Essential features of separated flows

Model II. Prosthetic Heart Valve Flow

Damaged mitral and/or aortic heart valves are causes for heart disease. These damaged valves are often replaced by prosthetic heart valves. Some of these have been shown in [35]. Flow characteristics of these prostheses are rather complicated. We need to consider viscous incompressible flow in a region containing immersed boundaries which move in the flow and exert forces on it. Mathematical modeling of these phenomena is extremely hard. However basic features of viscous flow, like flow separation, reattachment, generation of vortices, etc. are present here (Figures 5.2, f.3 and 5.4). Mueller and his colleagues [21] made some numerical studies on prosthetic heart valve flow. It is interesting to note that some of the conclusions drawn by Mueller [22] have been recently verified experimentally by Yoganathan, et al. [35]. In short these are that the present artificial heart valves cause (i) destruction of red-cells, (ii) thrombus formation, (iii) damage to the endothelial lining of the ascending aorta, and (iv) tissue overgrowth.

Let us now consider the model discussed in [20].

The Navier-Stokes equations and the equation of continuity in a cylindrical coordinate system (r, θ, z) for two-dimensional axisymmetric incompressible flow are:

$$\frac{\partial u}{\partial t} + u\frac{\partial u}{\partial r} + v\frac{\partial u}{\partial z} = -\frac{1}{\rho}\frac{\partial p}{\partial r} + \nu\left(\frac{\partial^2 u}{\partial r^2} + \frac{\partial^2 u}{\partial z^2} + \frac{1}{r}\frac{\partial u}{\partial r} - \frac{u}{r^2}\right) \tag{5.4}$$

$$\frac{\partial v}{\partial t} + u\frac{\partial v}{\partial r} + v\frac{\partial v}{\partial z} = -\frac{1}{\rho}\frac{\partial p}{\partial z} + \nu\left(\frac{\partial^2 v}{\partial r^2} + \frac{\partial^2 v}{\partial z^2} + \frac{1}{r}\frac{\partial v}{\partial r}\right) \tag{5.5}$$

and

$$\frac{\partial (ru)}{\partial r} + \frac{\partial (rv)}{\partial z} = 0 \tag{5.6}$$

where

u = velocity along the radial direction (r)

v = velocity along the axial direction (z)

ρ = density

p = pressure

and

ν = kinematic coefficient of viscosity.

If ζ is the vorticity for this flow,

$$\zeta = \left(\frac{\partial u}{\partial z} - \frac{\partial v}{\partial r}\right) . \tag{5.7}$$

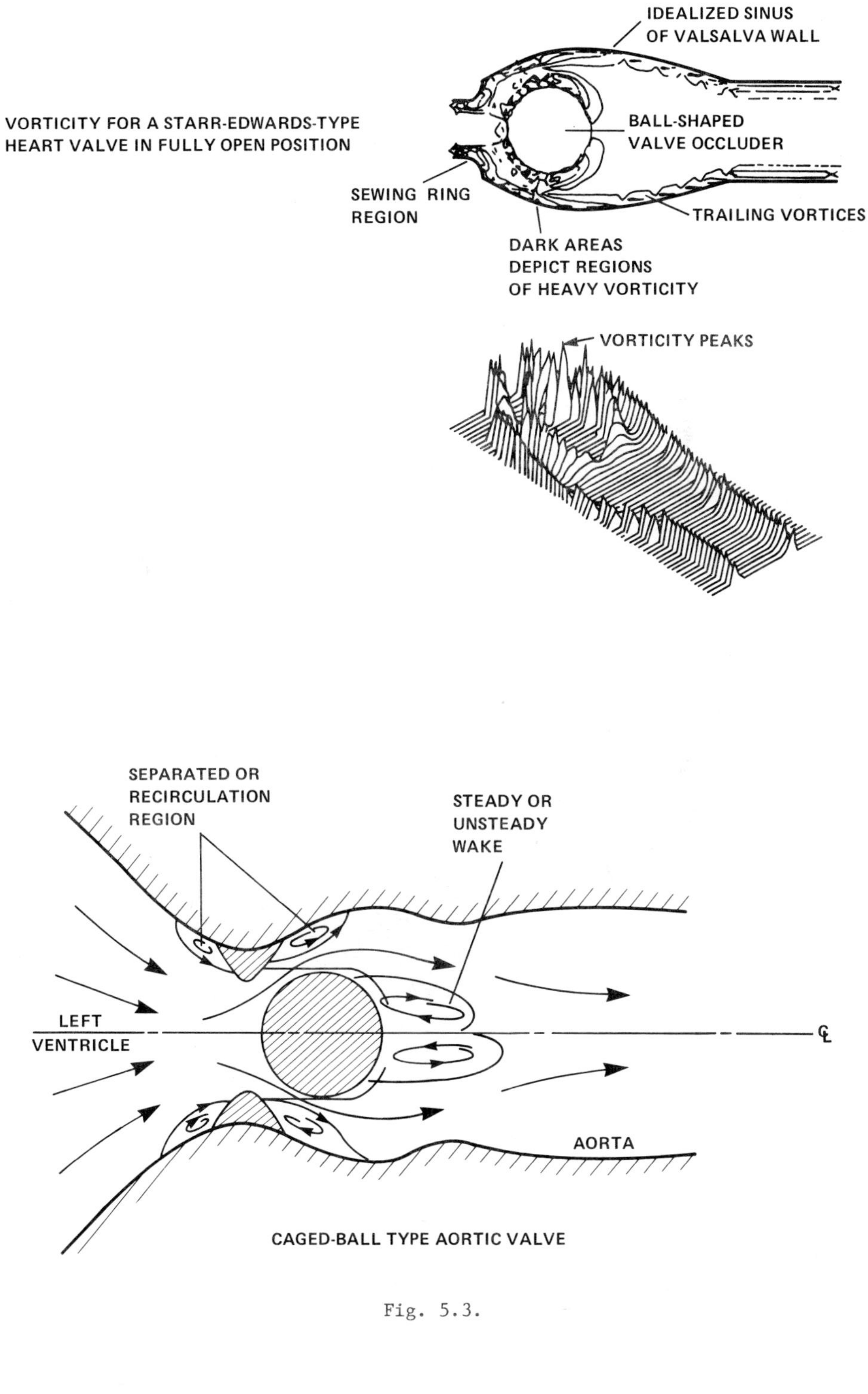
VORTICITY FOR A STARR-EDWARDS-TYPE HEART VALVE IN FULLY OPEN POSITION
IDEALIZED SINUS OF VALSALVA WALL
BALL-SHAPED VALVE OCCLUDER
SEWING RING REGION
TRAILING VORTICES
DARK AREAS DEPICT REGIONS OF HEAVY VORTICITY
VORTICITY PEAKS
SEPARATED OR RECIRCULATION REGION
STEADY OR UNSTEADY WAKE
LEFT VENTRICLE
℄
AORTA
CAGED-BALL TYPE AORTIC VALVE

Fig. 5.3.

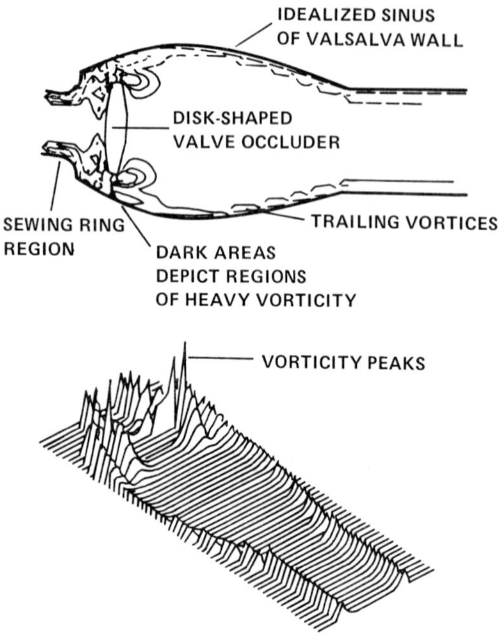

VORTICITY FOR A KAY-SHILEY-TYPE HEART VALVE IN FULLY OPEN POSITION

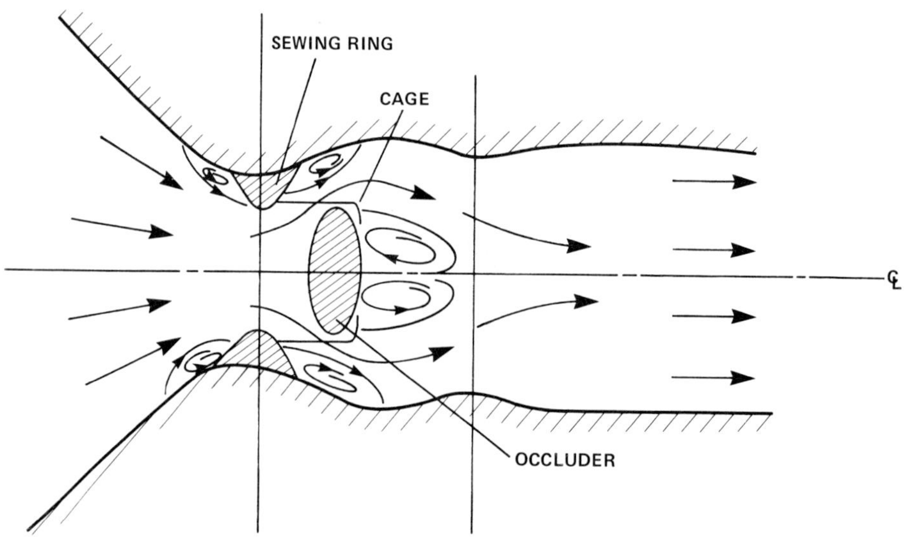

CAGED-DISK TYPE AORTIC VALVE

Fig. 5.4.

Eliminating p from (5.4) and (5.5) and using (5.7) we get:

$$\frac{\partial\zeta}{\partial t} + u\frac{\partial\zeta}{\partial r} + v\frac{\partial\zeta}{\partial z} - \frac{u\zeta}{r} = \nu\left(\frac{\partial^2\zeta}{\partial r^2} + \frac{\partial^2\zeta}{\partial z^2} + \frac{1}{r}\frac{\partial\zeta}{\partial r} - \frac{\zeta}{r^2}\right) . \quad (5.8)$$

If ψ is the stream-function defined by:

$$\frac{\partial\psi}{\partial r} = rv \quad \text{and} \quad \frac{\partial\psi}{\partial z} = -ru, \quad (5.9)$$

Equation (5.7) becomes:

$$\frac{\partial^2\psi}{\partial r^2} + \frac{\partial^2\psi}{\partial z^2} - \frac{1}{r}\frac{\partial\psi}{\partial r} = -r\zeta \quad (5.10)$$

The components of the stress tensor are:

$$\begin{aligned} \tau_{rr} &= -2\mu\frac{\partial u}{\partial r} \\ \tau_{\theta\theta} &= -2\mu\frac{u}{r} \\ \tau_{zz} &= -2\mu\frac{\partial v}{\partial z} \\ \tau_{zr} = \tau_{rz} &= -\mu\left(\frac{\partial v}{\partial r} + \frac{\partial u}{\partial z}\right) \end{aligned} \quad (5.11)$$

Equation (5.8) may be expressed as:

$$\frac{\partial\zeta}{\partial t} + \frac{\partial(u\zeta)}{\partial r} + \frac{\partial(v\zeta)}{\partial z} = \nu\left[\frac{\partial^2\zeta}{\partial r^2} + \frac{\partial^2\zeta}{\partial z^2} + \frac{1}{r}\frac{\partial\zeta}{\partial r} - \frac{\zeta}{r^2}\right] \quad (5.12)$$

These equations, subject to appropriate boundary conditions, were solved by Underwood and Mueller [21] by using an explicit upwind finite-difference scheme. These schemes, in general, have poor stability properties, and small values of time steps are required in order to obtain a meaningful solution. This was done by Mueller in his Equation (45). The method employed is known as "upwind differencing" [26]. This was used on (5.12) where $\partial\zeta/\partial t$ was approximated by a forward difference approximation. First order space derivatives representing advection terms were approximated by backward differences and second order space derivatives representing viscous diffusion terms were approximated by central differences. Thus effects of advection were felt only in the flow direction and those of diffusion were felt in all directions. A modification was done to the first order space derivatives by approximating them by forward differences at a nodal point in the finite difference field if the velocities in the neighborhood of that node are all negative. Obviously for this upwind differencing method information is advected into a cell only from those cells that are upwind of it. This is physically correct because transport property is advected only in the direction of the velocity. The basic problem is that any error, however small, originating in the computation at a nodal point in the upstream position will be directly transported to the cells in the downstream direction. Also, Mueller noted that: "The accuracy of these upwind differencing methods in resolving viscous effects usually deteriorates at high Reynolds number because they implicitly introduce an effect which resembles an artificial viscosity effect. Although this numerical viscosity is not unique to upwind differencing, it appears to be the predominant source of errors in these schemes." This definitely implies that some finite difference methods are needed to solve physiological models where computational errors will be small and computer-run-time and memory storage requirements will be also small. Mueller obtained good results validated by experimental data up to a Reynolds number equal to 500. Reynolds number for a laminar flow of blood is about 10^4. Implicit finite methods were applied by Dey [11] and Thompson [32] to solve Navier-Stokes equation for viscous flows at large Reynolds numbers, and no instabilities were found. In [11] the largest Reynolds number was 2×10^4 for time-dependent viscous flow past a circular cylinder. The method for solution was Gauss-Seidel iteration and computational time was large even when rate of convergence was accelerated by introducing SOR-scheme. Hence better numerical methods are required in computational fluid dynamics to solve nonlinear partial differential equations.

With these comments, I now intend to develop a new method which could be applied to solve nonlinear differential equations numerically.

A NUMERICAL METHOD FOR NONLINEAR SYSTEMS

Theoretically, Newton's method which has a quadratic rate of convergence should be used to solve nonlinear systems. But to execute the algorithm one needs to compute Jacobians for each iterate and at each iteration. Thus, to solve large nonlinear systems this method is not quite practical. Furthermore, in order to make the algorithm computationally effective in general, one should make an initial estimate for the root close to the solution. For problems, related to physical science, the nature of the solutions may be studied experimentally and/or mathematically. But for many problems in biological sciences, especially those mentioned in the first section, the nature of solutions of the mathematical models are hard to predict. Thus for numerical solution of these models, Newton's method was not used. There are several other methods [23] for nonlinear systems. But their computational effectiveness with regard to numerical solution of nonlinear differential equations are yet to be known.

Here we will describe a simple numerical scheme involving functional iterations and could be effectively used to solve nonlinear systems. The method is obtained by perturbing existing functional iterations and as such it is called a Perturbed Iterative Scheme or simply PIS.

The Basic Concept in PIS [8]

Let us consider an equation $f(x) = 0$, which has a root $x = x^* \in [a, b]$. If we write $f(x) = x - g(x)$, the equation becomes:

$$x = g(x) \tag{5.13}$$

and obviously,

$$x^* = g(x^*). \tag{5.14}$$

If we apply Picard's iteration to solve (5.13), then at some p^{th} iteration,

$$x^p = g(x^{p-1}) \tag{5.15}$$

where superscripts denote iterations and not exponents. Let us perturb this iterative scheme and write:

$$x^p = \omega^p + g(x^{p-1}) \tag{5.16}$$

where the perturbation parameter ω^p is yet to be known. This is done with the expectation that the convergence properties of (5.15) will be enhanced with the introduction of perturbation parameter ω^p.

In order to compute ω^p we make the following assumptions: (i) $g'(x) \neq 1$ and $g''(x)$ is bounded $\forall\ x \in [a, b]$ and $g(x^p) \in [a, b]\ \forall\ p$; (ii) ω^p's are small such that terms of the order $(\omega^p)^2$ may be neglected. Now, if (5.16) converges after p iterations, $x^* = x^p$. Then from (5.14),

$$\omega^p + g^{p-1} = g(\omega^p + g^{p-1}) \tag{5.17}$$

where $g^{p-1} = g(x^{p-1})$. Expanding the right side of (5.17) by Taylor's series and neglecting terms of the order $(\omega^p)^2$ by virtue of the above assumptions we get:

$$\omega^p = \{1 - g'(g^{p-1})\}^{-1}\{g(g^{p-1}) - g^{p-1}\}. \tag{5.18}$$

This defines an iterative scheme as follows: at some p^{th} iteration first compute ω^p using (5.18) and then add it to $g(x^{p-1})$ to compute the pth iterate x^p. Obviously the method fails if $g'(g^{p-1}) = 1$ at some iteration. We call it PIS (Perturbed Iterative Scheme). It may be seen that PIS is a combination of Picard's method and Newton's method. However an arbitrary combination of two existing methods may not be mathematically or computationally fruitful. PIS, on the other hand, has a development of its own and its mathematical properties have been investigated quite thoroughly ([7], [8], [9]).

It may be proved:

THEOREM 1. A necessary condition for convergence of (5.16) is that

$$\lim_{p\to\infty} |\omega^p| = 0. \tag{5.19}$$

THEOREM 2. If (i) $\forall$ y, z $\in$ [a, b],

$$|g(y) - g(z)| \leq L|y - z|, \quad 0 \leq L < 1$$

and (ii) $\forall$ p, $x^p \in [a, b]$, then (5.19) is a sufficient condition so that (5.16) will converge to $x^* \in [a, b]$ and furthermore x^* is the unique root in [a, b].

THEOREM 3. PIS given by (5.16) has a quadratic rate of convergence.

These theorems have been proved in [8] and are not repeated here.

Computationally, it has been found that PIS has a global convergence property. For equations with multiple roots PIS is not quite effective, because one cannot predict to which root it will converge. In a computer code, we always use (5.19) as our convergence criterion. For equations with one unknown, PIS is as good as Newton's method or Aitken's δ^2-process.

A SIMPLE EXAMPLE. Solve $x = 1/\sqrt{x}$. This equation has only one real root, x = 1. Here $g(x) = 1/\sqrt{x}$, $g'(x) = -1/2x^{3/2}$; $g^{p-1} = (x^{p-1})^{-1/2}$; $\omega^p = \{1 - g'(g^{p-1})\}^{-1}\cdot\{g(g^{p-1}) - g^{p-1}\}$; $x^p = \omega^p + g^{p-1}$. Convergence criterion is $|\omega^p| < 0.001$. With an initial estimate $x^0 = 0.0001$, PIS converged to the root within five iterations.

PIS for Nonlinear Systems [7]

Let us consider a system of nonlinear equations given by

$$f_i(x_1, x_2, \ldots, x_n) = 0 \tag{5.20}$$

(i = 1, 2, ..., n) in a domain $D \subset R^n$, where R^n is a real n-dimensional space. Let $x = (x_1\ x_2\ \ldots\ x_n)^T \in D$ and $F: D \subset R^n \to R^n$. Then (5.20) may be expressed as:

$$F(x) = 0. \tag{5.21}$$

We assume that $x = x^* \in D$ is a root. Then, $F(x^*) = 0$. To apply nonlinear Gauss-Seidel iterations to solve this system we may express (5.20) as:

$$x_i = g_i(x_1, x_2, \ldots, x_n) \tag{5.22}$$

$$i = 1, 2, \ldots, n$$

Assuming $x^0 \in D$ to be an initial estimate of x^* at some p^{th} iteration we have

$$x_i^p = g_i^p \tag{5.23}$$

where

$$g_i^{\,p} = g_i(x_1^{\,p}, x_2^{\,p}, \ldots, x_{i-1}^{\,p}, x_i^{\,p-1}, \ldots, x_n^{\,p-1}) \tag{5.24}$$

Assuming $x^p \in D$ and a mapping $G: D \times D \subset R^n \times R^n \to D$ (5.24) may be expressed as:

$$x^p = G(x^p, x^{p-1}) \tag{5.25}$$

Then, $x^* = G(x^*, x^*)$ and PIS is

$$x^p = \omega^p + G(x^p, x^{p-1}). \tag{5.26}$$

In the element form this scheme is:

$$x_i^p = \omega_i^p + g_i^p \tag{5.27}$$

where g_i^p is given by (5.24).

To compute ω_i^p we make the following assumptions: $\forall$ p and $x^p \in D$ (i) $(\partial g_i/\partial x_i^p) \neq 1$, (ii) $\partial^2 g_i/\partial x_i^{p^2}$ is bounded, and (iii) ω_i^p's are small such that terms of the order $(\omega_i^p)^2$ may be neglected. If (5.27) converges after $p - 1$ iterations, $x_i^{p-1} = x_i^p = x_i^*$ $\forall$ $i = 1, 2, \ldots, n$. This gives

$$\omega_i^p + g_i^p = g_i(x_1^p, x_2^p, \ldots, x_{i-1}^p, \omega_i^p + g_i^p, x_{i+1}^{p-1}, \ldots, x_n^{p-1}).$$

Expanding the right side of the above equation by Taylor's series and neglecting terms of the order $(\omega_i^p)^2$ by virtue of our assumptions, we get:

$$\omega_i^p = \{g_i(x_1^p, \ldots, x_{i-1}^p, g_i^p, x_{i+1}^{p-1}, \ldots, x_n^{p-1}) - g_i^p\}\cdot\{1 - \partial_i^p g_i\}^{-1} \tag{5.28}$$

where

$$\partial_i^p g_i = (\partial g_i/\partial x_i)_{x_1^p, \ldots, x_{i-1}^p, g_i^p, x_{i+1}^{p-1}, \ldots, x_n^{p-1}}$$

Thus, the perturbation parameters, ω_i^p's are computed in terms of quantities known apriori.

The algorithm for computation may now be stated as follows:

Step #1. Express the given nonlinear system in the form (5.22).

Step #2. Make an initial estimate for the root x^* as

$$x^0 = (x_1^0\ x_2^0 \ldots x_n^0)^T.$$

Step #3. For $p = 1, 2, \ldots$ and $i = 1, 2, \ldots, n$ compute the following:

(a) g_i^p using (5.24)

(b) $g_i(x_1^p, \ldots, x_{i-1}^p, g_i^p, x_{i+1}^{p-1}, \ldots, x_n^{p-1})$

(c) ω_i^p using (5.28)

(d) $x_i^p = \omega_i^p + g_i^p$

(e) $W = \max_i |\omega_i^p|$

Step #4. If at some p, $W < \varepsilon$, where ε is positive and arbitrarily small, iterations converge.

An Application: Consider the system:

$$x = \frac{x}{x^2 + y^2} - e^{-x} \cos y + 1.712173$$

$$y = \frac{y}{x^2 + y^2} + e^{-x} \sin y + 3.211671.$$

The solution is: $x = 1.9744$, $y = 3.3967$. Here

$$g_1(x, y) = x/(x^2 + y^2) - e^{-x} \cos y + 1.712173,$$

$$\partial g_1/\partial x = (y^2 - x^2)/(x^2 + y^2) + e^{-x} \cos y,$$

$$g_2(x, y) = y/(x^2 + y^2) + e^{-x} \sin y + 3.211671,$$

$$\partial g_2/\partial y = (x^2 - y^2)/(x^2 + y^2) + e^{-x} \cos y.$$

Now, following the algorithm: Make an initial guess, $x^0 = -4.5$, $y^0 = 65.0$, and at some p^{th} iteration compute:

$$g_1^p = g_1(x^{p-1}, y^{p-1})$$

$$= \frac{x^{p-1}}{(x^{p-1})^2 + (y^{p-1})^2} - e^{-(x^{p-1})} \cos (y^{p-1}) + 1.712173$$

$$\left[\frac{\partial g_1}{\partial x_1^p}\right]_{g_1^p,\, y^{p-1}} = \frac{(y^{p-1})^2 - (g_1^p)^2}{(g_1^p)^2 + (y^{p-1})^2} + e^{-(g_1^p)} \cos (y^{p-1})$$

$$\omega_1^p = \frac{g_1(g_1^p, y^{p-1}) - g_1^p}{1 - \left[\frac{\partial g_1}{\partial x_1^p}\right]_{g_1^p,\, y^{p-1}}}$$

$$x^p = \omega_1^p + g_1^p$$

$$g_2^p = g_2(x^p, y^{p-1})$$

$$= \frac{y^{p-1}}{(x^p)^2 + (y^{p-1})^2} + e^{-(x^p)} \sin (y^{p-1}) + 3.211671$$

$$\left[\frac{\partial g_2}{\partial y}\right]_{x^p,\, g_2^p} = \frac{(x^p)^2 - (g_2^p)^2}{(x^p)^2 + (g_2^p)^2} + e^{-(x^p)} \cos (g_2^p)$$

$$\omega_2^p = \frac{g_2(x^p, g_2^p) - g_2^p}{1 - \left[\frac{\partial g_2}{\partial y}\right]_{x^p,\, g_2^p}}$$

$$y^p = \omega_2^p + g_2^p .$$

Convergence criterion is at some p^{th} iteration:

$$\max_{i=1,2} (|\omega_i^p|) < \varepsilon.$$

When we choose $\varepsilon = 10^{-3}$, this method converges within 3 iterations.

PIS for a Two-dimensional Finite-difference System.

Let us consider a two dimensional finite difference field described by I number of points along the x-axis and J number of points along the y-axis. In total, there are IJ number of points in the field. Let u_{ij} be the unknowns to be computed from the difference equations:

$$u_{ij} = f_{ij}(u_{11}, \ldots, u_{ij}, \ldots, u_{IJ}) \tag{5.29}$$

where $i = 1, 2, \ldots, I;\ j = 1, 2, \ldots, J$.

Let D be the domain where (5.29) has a solution given by

$$u_{ij}^* = f_{ij}(u_{11}^*, \ldots, u_{ij}^*, \ldots, u_{IJ}^*) \tag{5.30}$$

$$(i = 1, 2, \ldots, I;\ j = 1, 2, \ldots, J).$$

Let $U_i = (u_{i1}, u_{i2}, \ldots, u_{iJ})^T$, where $i = 1, 2, \ldots, I$.

The nonlinear system may now be expressed as:

$$U_i = F_i(U_1, U_2, \ldots, U_I) \tag{5.31}$$

$$(i = 1, 2, \ldots, I);$$

where

$$F_i = (f_{i1}, f_{i2}, \dots, f_{iJ})^T.$$

The solution is now:

$$U_i^* = (u_{i1}^*, u_{i2}^*, \dots, u_{iJ}^*)^T \tag{5.32}$$

$(i = 1, 2, \dots, I)$.

If at some p^{th} iteration

$$F_i^p = F_i(U_1^p, U_2^p, \dots, U_{i-1}^p, U_i^{p-1}, \dots, U_I^{p-1})$$

then PIS, at the pth iteration is:

$$U_i^p = W_i^p + F_i^p \tag{5.33}$$

$(i = 1, 2, \dots, I)$.

The perturbation parameters $W_i = (\omega_{i1}^p, \omega_{i2}^p, \dots, \omega_{iJ}^p)^T$ are yet to be computed. This iterative scheme shows that iterations proceed columnwise as follows:

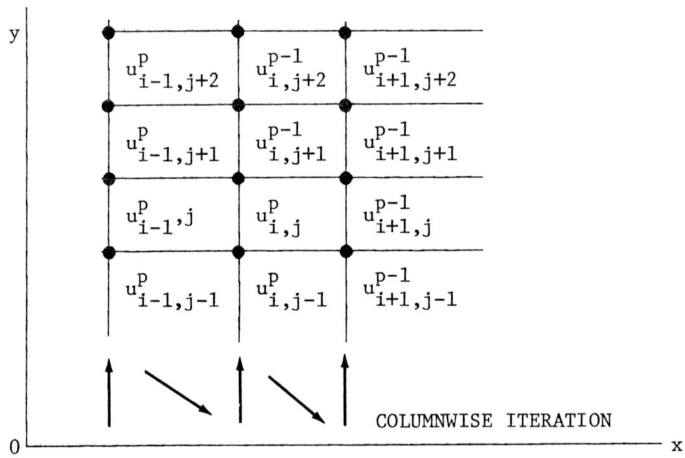

This shows that when we are computing u_{ij} at the pth iteration, values of u_{mj}, $(m = 1, 2, \dots, i - 1$, and $j = 1, 2, \dots, J)$ are already known at the pth iteration, values of u_{mj} $(m = i + 1, i + 2, \dots, I$, and $j = 1, 2, \dots, J)$ are known only at the $(p - 1)^{th}$ iteration, values of u_{in}, $(n = 1, 2, \dots, j - 1)$ are known at the pth iteration and values of u_{in}, $(n = j + 1, j + 2, \dots, J)$ are known at the $(p - 1)^{th}$ iteration. We may call this procedure a columnwise iteration. If we define $U_j = (u_{1j}\ u_{2j} \dots u_{Ij})^T$, $j = 1, 2, \dots, J$, then (5.31) becomes

$$U_j = F_j(U_1, U_2, \ldots, U_J), \qquad (j = 1, 2, \ldots, J),$$

where

$$F_j = (f_{1j}\ f_{2j}\ \cdots\ f_{Ij})^T.$$

In that case PIS will follow a rowwise iteration.

From (5.33), if PIS converges after (p - 1) iterations, $U_i^{p-1} = U_i^p = U_i^*$ for all i = 1, 2, ..., I. Thus in the element form,

$$\omega_{ij}^p + f_{ij}^p = f_{ij}(U_1^p, \ldots, U_{i-1}^p; u_i^p, \ldots, u_{ij-1}^p, \omega_{ij}^p + f_{ij}^p, u_{ij+1}^{p-1}, \ldots, u_{iJ}^{p-1}; U_{i+1}^{p-1}, \ldots, U_I^{p-1}) \tag{5.34}$$

where

$$f_{ij}^p = f_{ij}(U_1^p, U_2^p, \ldots, U_{i-1}^p; u_i^p, \ldots, u_{ij-1}^p, u_{ij}^{p-1}, \ldots, u_{iJ}^{p-1}; U_{i+1}^{p-1}, \ldots, U_I^{p-1}). \tag{5.35}$$

Let

$$f_{ij}(f_{ij}^p) = f_{ij}(U_1^p, \ldots, U_{i-1}^p; u_{i1}^p, \ldots, u_{ij-1}^p, f_{ij}^p, u_{ij+1}^{p-1}, \ldots, u_{iJ}^{p-1}, U_{i+1}^{p-1}, \ldots, U_I^{p-1}) \tag{5.36}$$

If we assume that ω_{ij}^p's are small, such that terms of the order $(\omega_{ij}^p)^2$ may be neglected and also

$$\frac{\partial f_{ij}}{\partial u_{ij}} \neq 1$$

where f_{ij}'s are continuously differentiable with respect to u_{ij}'s at least three times, then (using the notation in Eq. 5.36), we get, by Taylor's expansion:

$$\omega_{ij}^p = \frac{f_{ij}(f_{ij}^p) - f_{ij}^p}{1 - \left[\dfrac{\partial f_{ij}}{\partial u_{ij}}\right]_{U_1^p,\ldots,U_{i-1}^p,Q_i^p,U_{i+1}^{p-1},\ldots,U_I^{p-1}}} \tag{5.37}$$

$$(i = 1, 2, \ldots, I;\ j = 1, 2, \ldots, J);$$

where

$$Q^p = (u_{i1}^p\ u_{i2}^p\ \cdots\ u_{ij-1}^p\ f_{ij}^p\ u_{ij+1}^{p-1}\ \cdots\ u_{ij}^{p-1})^T.$$

The algorithm may now be expressed as follows:

At some p^{th} iteration, for each i = 1, 2, ..., I and j = 1, 2, ..., J compute:

(i) f^p_{ij} from (5.35)

(ii) ω^p_{ij} from (5.37) and

(iii) $u^p_{ij} = \omega^p_{ij} + f^p_{ij}$.

If

$$\max_{ij} |\omega^p_{ij}| < \varepsilon$$

where ε is positive and arbitrarily small, iterations converge.

For most time dependent fluid flows, it is advisable that ε should be chosen very small. A value of ε, equal to 10^{-10} or less is recommended.

Some detailed describtion of mathematical properties of PIS may be found in [7], [8], and [9].

For coupled nonlinear systems, some modifications of PIS may be done as follows:

$$u_{ij} = f_{ij}(u_{11}, \ldots, u_{IJ}; v_{11}, \ldots, v_{IJ}) \tag{5.38a}$$

$$v_{ij} = g_{ij}(u_{11}, \ldots, u_{IJ}; v_{11}, \ldots, v_{IJ}) \tag{5.38b}$$

where $i = 1, 2, \ldots, I;\ j = 1, 2, \ldots, J$.

There are altogether 2(IJ) number of equations. Depending upon the physical nature of the problem, one may use (i) columnwise iterations for u_{ij} and rowwise iterations for v_{ij} and vice-versa or (ii) rowwise iterations for both u_{ij} and v_{ij} or (iii) columnwise iterations for both u_{ij} and v_{ij}.

In case of columnwise (or rowwise) iterations for both u_{ij} and v_{ij}, we may now modify PIS to compute u^p_{ij} and v^p_{ij} simultaneously (and not first compute u^p_{ij} and then compute v^p_{ij} as before).

Let u^p_{ij} and v^p_{ij} be given by:

$$u^p_{ij} = \alpha^p_{ij} + f^p_{ij} \tag{5.39a}$$

$$v^p_{ij} = \beta^p_{ij} + g^p_{ij} \tag{5.39b}$$

$$(i = 1, 2, \ldots, I;\ j = 1, 2, \ldots, J)$$

where α^p_{ij} and β^p_{ij} are perturbation parameters and

$$f^p_{ij} = f_{ij}(U^p_1, U^p_2, \ldots, U^p_{i-1}, u^p_{i1}, \ldots, u^p_{ij-1}, u^{p-1}_{ij}, \ldots, u^{p-1}_{iJ},$$

$$U^{p-1}_{i+1}, \ldots, U^{p-1}_I; V^p_1, V^p_2, \ldots, V^p_{i-1}, v^p_{i1}, \ldots, v^p_{ij-1},$$

$$v^{p-1}_{ij}, \ldots, v^{p-1}_{iJ}, V^{p-1}_{i+1}, \ldots, V^{p-1}_{I})$$

g^p_{ij} = value of g_{ij} at the same point where f^p_{ij} is evaluated, and

$$U_i = (u_{i1}\ u_{i2} \cdots u_{iJ})^T$$

$$V_i = (v_{i1}\ v_{i2} \cdots v_{iJ})^T$$

$(i = 1, 2, \ldots, I)$.

If this modified PIS converges after $(p - 1)$ iterations:

$$\begin{aligned}\alpha^p_{ij} + f^p_{ij} = f_{ij}(&U^p_1, U^p_2, \ldots, U^p_{i-1}, u^p_{i1}, \ldots, u^p_{ij-1}, \alpha^p_{ij} + f^p_{ij}\\ &u^{p-1}_{ij+1}, \ldots, u^{p-1}_{iJ}, U^{p-1}_{i+1}, \ldots, U^{p-1}_{I}; V^p_1, V^p_2, \ldots, V^p_{i-1}\\ &v^p_{i1}, \ldots, v^p_{ij-1}, \beta^p_{ij} + g^p_{ij}, v^{p-1}_{ij+1}, \ldots, v^{p-1}_{iJ};\\ &V^{p-1}_{i+1}, \ldots, V^{p-1}_{I})\end{aligned}$$

$$\begin{aligned}\beta^p_{ij} + g^p_{ij} = g_{ij}(&U^p_1, U^p_2, \ldots, U^p_{i-1}, u^p_{i1}, \ldots, u^p_{ij-1}, \alpha^p_{ij} + f^p_{ij},\\ &u^{p-1}_{ij+1}, \ldots, u^{p-1}_{iJ}, U^{p-1}_{i+1}, \ldots, U^{p-1}_{I}; V^p_1, V^p_2, \ldots, V^p_{i-1},\\ &v^p_{i1}, \ldots, v^p_{ij-1}, \beta^p_{ij} + g^p_{ij}, v^{p-1}_{ij+1}, \ldots, v^{p-1}_{iJ};\\ &V^{p-1}_{i+1}, \ldots, V^{p-1}_{I})\end{aligned}$$

$(i = 1, 2, \ldots, I;\ j = 1, 2, \ldots, J)$.

To compute α^p_{ij} and β^p_{ij}, we use Taylor's theorem as before and get:

$$\left(1 - \frac{\partial f_{ij}}{\partial u_{ij}}\right)\alpha^p_{ij} + \left(-\frac{\partial f_{ij}}{\partial v_{ij}}\right)\beta^p_{ij} = f_{ij}(f^p_{ij}, g^p_{ij}) - f^p_{ij} \tag{5.40a}$$

$$\left(-\frac{\partial g_{ij}}{\partial u_{ij}}\right)\alpha^p_{ij} + \left(1 - \frac{\partial g_{ij}}{\partial v_{ij}}\right)\beta^p_{ij} = g_{ij}(f^p_{ij}, g^p_{ij}) - g^p_{ij} \tag{5.40b}$$

where

$$\begin{aligned}f_{ij}(f^p_{ij}, g^p_{ij}) = f_{ij}(&U^p_1, \ldots, U^p_{i-1}; u^p_{i1}, \ldots, u^p_{ij-1}, f^p_{ij},\\ &u^{p-1}_{ij+1}, \ldots, u^{p-1}_{iJ}; U^{p-1}_{i+1}, \ldots, U^{p-1}_{I}; V^p_1, \ldots, V^p_{i-1};\\ &v^p_{i1}, \ldots, v^p_{ij-1}, g^p_{ij}, v^{p-1}_{ij+1}, \ldots, v^{p-1}_{iJ}; V^{p-1}_{i+1}, \ldots, V^{p-1}_{I})\end{aligned}$$

$g_{ij}(f^p_{ij}, g^p_{ij})$ = value of g_{ij} at the same point as that for $f_{ij}(f^p_{ij}, g^p_{ij})$, and all the partial derivatives appearing in Equations (5.40a) and (5.40b) are evaluated at the same point as that for $f_{ij}(f^p_{ij}, g^p_{ij})$.

The equations (5.40a) and (5.40b) are solvable for α^p_{ij} and β^p_{ij}, provided the Jacobian of the system

$$1 - \left(\frac{\partial f_{ij}}{\partial u_{ij}} + \frac{\partial g_{ij}}{\partial v_{ij}}\right) + \left(\frac{\partial f_{ij}}{\partial v_{ij}}\frac{\partial g_{ij}}{\partial v_{ij}} - \frac{\partial f_{ij}}{\partial v_{ij}}\frac{\partial g_{ij}}{\partial u_{ij}}\right) \neq 0.$$

It is obvious that we no longer need the following restrictions:

$$\frac{\partial f_{ij}}{\partial u_{ij}} \neq 1, \qquad \frac{\partial g_{ij}}{\partial v_{ij}} \neq 1.$$

This modified PIS introduces the concept of multiple-point-iterations (where at each iteration, multiple iterates are to be computed) as opposed to point wise iterations in the regular PIS [7]. However, if several iterates are computed simultaneously at the same iteration, computation of Jacobians will be complicated and thus the inherent simplicity of the algorithm of PIS will be lost.

We may now make certain comments regarding the application of PIS ([7], [8]):

1. For finite different schemes with higher order accuracy, standard implicit methods are not found to be suitable, simply because matrices are no longer tridiagonal. In such cases PIS is applicable.

2. It may be noticed that PIS is essentially a combination of nonlinear Gauss-Seidel iterations and Lieberstien's method.

3. PIS has been successfully combined with relaxation parameters to generate faster rate of convergence [10]. However, computations of optimized relaxation parameters to obtain the fastest rate of convergence are yet to be done.

APPLICATIONS OF PIS TO NONLINEAR FINITE DIFFERENCE SYSTEMS

Most mathematical models representing flows of fluids in physics or in biomechanics deal with nonlinear partial differential equations. And, in general, analytical solutions of these models are yet to be found. Hence numerical methods are used to obtain some approximate solution. In this area, finite differences are widely applied to approximate partial differential equations by partial difference equations and solve these equations using digital computers. In biomechanics such examples are abundant. PIS has been developed in order to make it an application-oriented method. Hence to check its validity PIS must be applied to nonlinear models whose solutions are known exactly or approximately. In [10] several nonlinear ordinary differential equations representing two-point boundary value problems were approximated by consistent finite difference equations and were solved by PIS. The results, which were compared against existing analytical and/or numerical solutions, seem to be quite encouraging. Here we intend to present PIS solutions of two finite difference models representing two phenomena of physics of fluids. Very similar equations appeared in several problems in biomechanics [5], [16], [18], and [29]. Solutions of these two problems are known approximately.

Model I. Burgers' Equation [9]

Let us consider the nonlinear Burgers' equation:

$$\frac{\partial u}{\partial t} + u\frac{\partial u}{\partial x} = \nu\frac{\partial^2 u}{\partial x^2} \tag{5.41}$$

with the following initial-boundary conditions:

$$u(x, 0) = \sin\pi x, \qquad u(0, t) = u(1, t) = 0 \tag{5.42}$$

The Equation (5.41) is a second order nonlinear parabolic partial differential equation, and thus it retains certain intrinsic properties of Navier-Stokes equation which represents flow of viscous fluid.

The above model has a unique analytical solution given by:

$$u(x, t) = 2\pi\nu S_1(x, t)/S_2(x, t) \tag{5.43}$$

where

$$S_1(x, t) = \sum_{n=1}^{\infty} \exp(-n^2\pi^2\nu t)A_n\cdot n\cdot\sin(n\pi x), \tag{5.44}$$

$$S_2(x, t) = \sum_{n=0}^{\infty} \exp(-n^2\pi^2\nu t)A_n\cdot\cos(n\pi x), \tag{5.45}$$

$$A_0 = \int_0^1 \exp[(\cos\pi x - 1)/2\pi\nu]dx, \tag{5.46}$$

and

$$A_n = 2\int_0^1 \exp[(\cos\pi x - 1)/2\pi\nu]\cos(n\pi x)dx. \tag{5.47}$$

Assuming $\nu \neq 0$, it can be seen that the integrals (5.46) and (5.47) exist for all n and $\lim_{n\to\infty} A_n = 0$. As $\nu \to 0$, the integrals become increasingly difficult to compute and hence the analytical solution (5.43) becomes extremely hard to compute. The same is true for its numerical solution. A total failure of the explicit finite difference solution of this model was recorded by Ames [1] (Figures 2-10) for ν = 0.0001 and Δt = time step = 0.002 sec. (Small values of ν (viscosity) imply large values of the free stream Reynold's number.) This possibly will explain why Fromm [12] encountered instabilities for explicit finite difference solution of Navier-Stokes' equation

$$\frac{\partial\zeta}{\partial t} + u\,\frac{\partial\zeta}{\partial x} + v\,\frac{\partial\zeta}{\partial y} = \nu\nabla^2\zeta \tag{5.48}$$

$$\nabla^2\psi = -\zeta \tag{5.49}$$

where ζ = vosticity, ψ = stream function and u, v = velocity components along the positive directions of the x and y axes respectively. In Biomechanics explicit finite difference methods were used by several authors [15], [16], [20], and [29] to compute blood circulatory problems. For small Reynold's numbers, results as found by Lee and Fung [15] should be reasonably accurate. However for prosthetic heart valve flow of blood at a Reynold's number 10^3 [20] some experimental validation of the results is necessary.

We will now present the solution of an implicit finite difference model of Burgers' equation by PIS for various values of ν. Approximating the time derivative by a two-point backward difference formula and space derivatives by central differences an implicit finite difference analog corresponding to (5.41) is:

$$\frac{U_i^n - U_i^{n-1}}{k} + U_i^n\,\frac{U_{i+1}^n - U_{i-1}^n}{2h} = \frac{U_{i+1}^n - 2U_i^n + U_{i-1}^n}{h^2} \tag{5.50}$$

$$i = 1, 2, \ldots, I.$$

The initial/boundary conditions are:

$$U_i^0 = \sin(i\pi h) \qquad i = 1, 2, \ldots, I \tag{5.51}$$

$$U_0^n = U_{I+1}^n = 0 \qquad n = 0, 1, 2, \ldots$$

where k = time step = Δt, h = mesh size = Δx, $U_i^n = U(x_i, t_n)$ = net function corresponding to $u(x_i, t_n)$. The Equation (5.50) may be expressed, at each t_n, as:

$$x_i = g_i(x_1, x_2, \ldots, x_I) + c$$

where, $x_i = U_i^n$, $c = h^2 a U_i^{n-1}$, with $a = (h^2 + 2\nu k)^{-1}$, and

$$g_i(x_1, \ldots, x_n) = 0.5hkax_i(x_{i-1} - x_{i+1}) + \nu ka(x_{i+1} + x_{i-1}).$$

Thus,

$$\frac{\partial g_i}{\partial x_i} = 0.5\ hka(x_{i-1} - x_{i+1}).$$

Application of PIS now requires at each iteration p = 1, 2, ... for i = 1, 2, ..., I

$$g_i^p = 0.5hkax_i^{p-1}(x_{i-1}^p - x_{i+1}^{p-1}) + \nu ka(x_{i+1}^{p-1} + x_{i-1}^p)$$

$$g_i(x_i^p, \ldots, x_{i-1}^p, g_i^p, x_{i+1}^{p-1}, \ldots, x_I^{p-1})$$
$$= 0.5hkag_i^p(x_{i-1}^p - x_{i+1}^{p-1}) + ka(x_{i+1}^{p-1} + x_{i-1}^p)$$

$$\left[\frac{\partial g_i}{\partial x_i}\right]_{x_1^p, \ldots, x_{i-1}^p, g_i^p, x_{i+1}^{p-1}, \ldots, x_I^{p-1}} = 0.5hka(x_{i-1}^p - x_{i+1}^{p-1})$$

$$\omega_i^k = \frac{g_i(x_i^p, \ldots, x_{i-1}^p, g_i^p, x_{i+1}^{p-1}, \ldots, x_I^{p-1}) - g_i^p}{1 - (\partial g_i/\partial x_i)_{x_1^p, \ldots, x_{i-1}^p, g_i^p, x_{i+1}^{p-1}, \ldots, x_I^{p-1}}}$$

$$x_i^p = \omega_i^p + g_i^p$$

at each t_n.

PIS converges if at some p^{th} iteration

$$\max_i |\omega_i^p| < \varepsilon.$$

It should be noted that ε must be chosen to be small. In the computer code we chose $\varepsilon = 10^{-10}$. In some cases this might increase rather unnecessarily, the number of iterations for convergence and hence computer run-time is also increased. But in general, since the rate of convergence of PIS is quadratic, small ε should not significantly increase the number of iterations. For Burgers' equation, computer run-time was rather low. With various time steps Δt = 0.0025, 0.005, and 0.01 seconds within no more than 5 iterations PIS converged. In the Tables I and II some comparisons between exact and numerical solutions by PIS for various values of ν are given. In the Figures 5.5 and 5.6 we have found a sequence of velocity profiles (obtained by PIS) and development of sharp discontinuities near x = 1.0 as predicted by mathematical properties of the model. The profiles corresponding to ν = 0.0001 have been presented in [9].

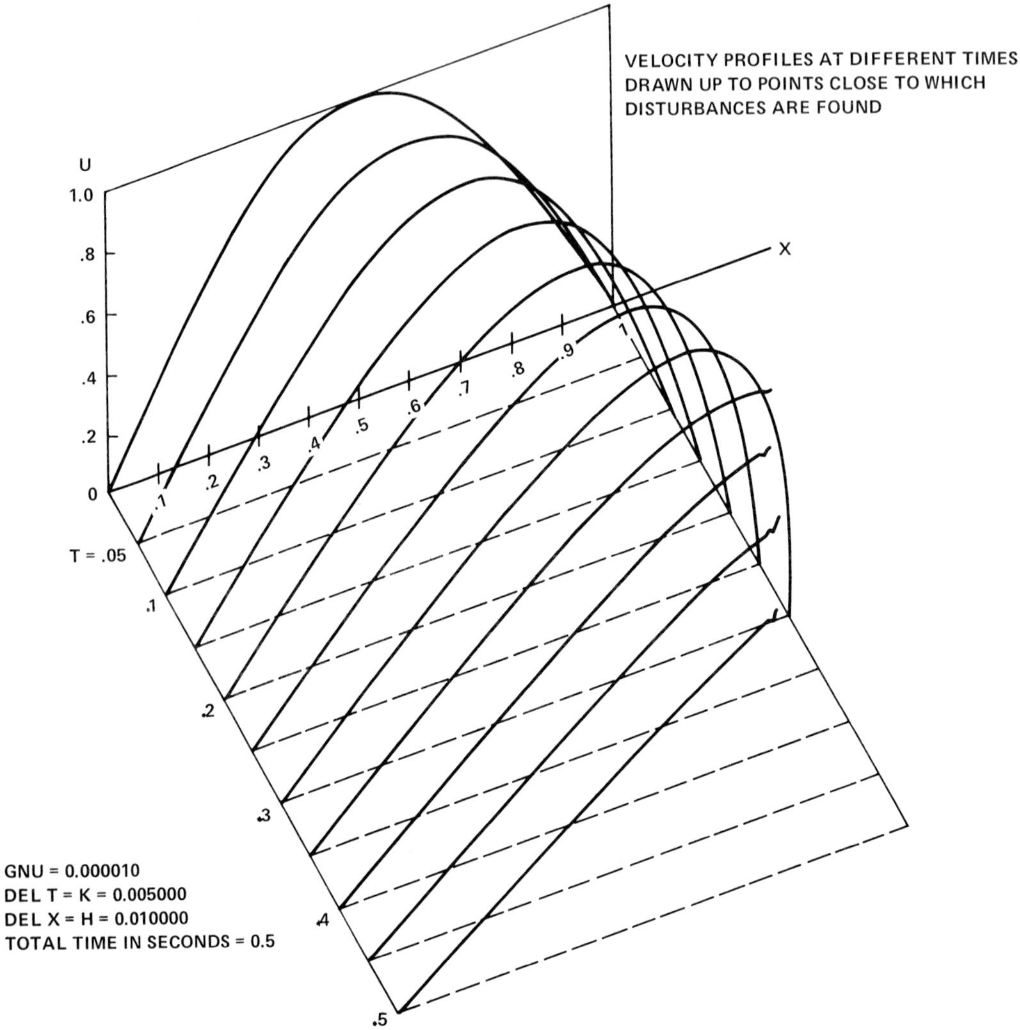
VELOCITY PROFILES AT DIFFERENT TIMES
DRAWN UP TO POINTS CLOSE TO WHICH
DISTURBANCES ARE FOUND
U
1.0
.8
.6
.4
.2
0
X
.1
.2
.3
.4
.5
.6
.7
.8
.9
1
T = .05
.1
.2
.3
.4
.5
GNU = 0.000010
DEL T = K = 0.005000
DEL X = H = 0.010000
TOTAL TIME IN SECONDS = 0.5

Fig. 5.5.

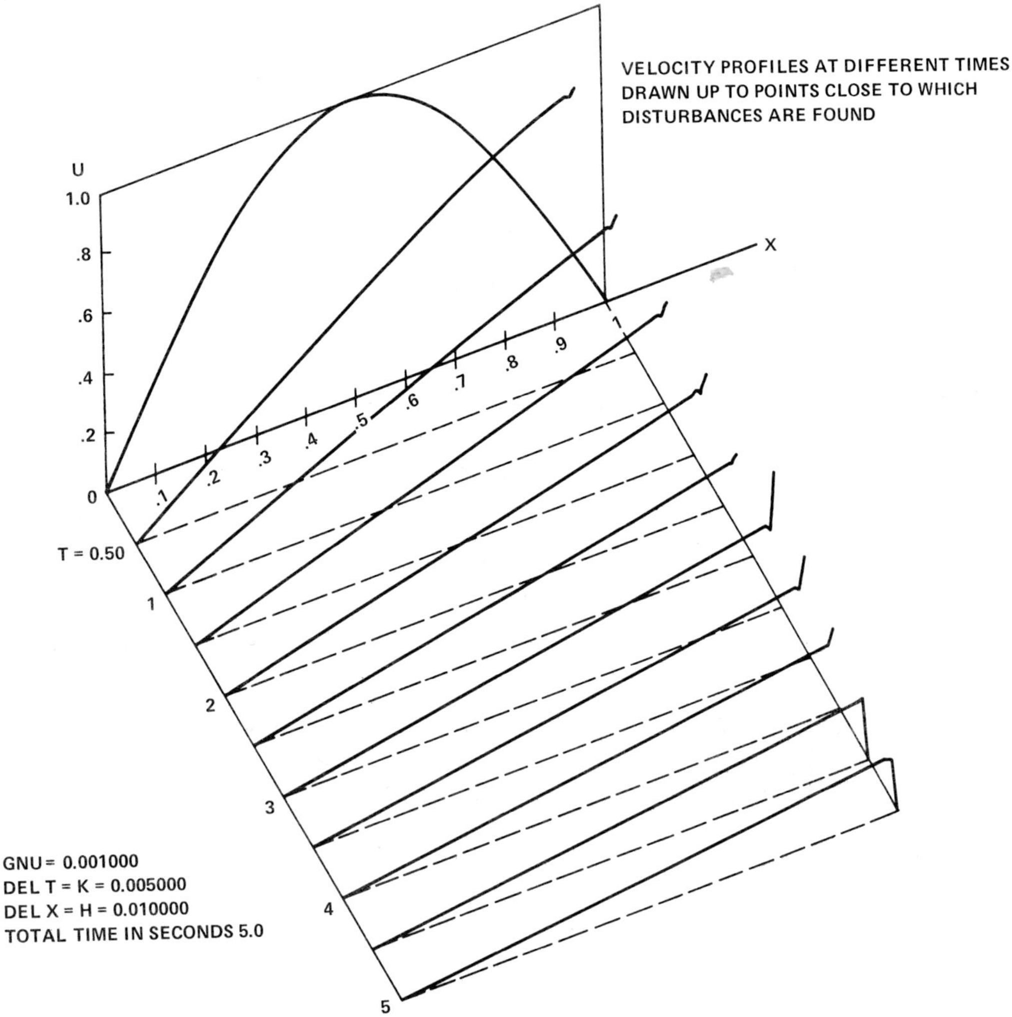
VELOCITY PROFILES AT DIFFERENT TIMES DRAWN UP TO POINTS CLOSE TO WHICH DISTURBANCES ARE FOUND
U
1.0
.8
.6
.4
.2
0
X
.1
.2
.3
.4
.5
.6
.7
.8
.9
1
T = 0.50
1
2
3
4
5
GNU = 0.001000
DEL T = K = 0.005000
DEL X = H = 0.010000
TOTAL TIME IN SECONDS 5.0

Fig. 5.6.

BURGERS' EQUATION

Table I

Values of U_i^n Computed by Analytical Method and by PIS at $x_i = 0.5$ ft. and $t_n = 0.05$ sec. for different ν and Δt.

νft.^{2}/sec.	Exact	PIS Δt=0.0025	PIS Δt=0.005	PIS Δt=0.01
1.0	0.60902	0.61287	0.61650	0.62347
0.1	0.94229	0.94219	0.94202	0.94170
0.01	0.98337	0.98286	0.98237	0.98144
0.001†	Failed	0.98700	0.98648	0.98548

Table II

Values of U_i^n Computed by Analytical Method and by PIS at $x_i = 0.5$ ft. and $t_n = 0.25$ sec. for different ν and Δt.

νft.^{2}/sec.	Exact	PIS Δt=0.0025	PIS Δt=0.005	PIS Δt=0.01
1.0	0.08453	0.08781	0.09038	0.09549
0.1	0.69994	0.70072	0.70138	0.70269
0.01	0.79675	0.79707	0.79733	0.79785
0.001†	Failed	0.80550	0.80573	0.80619

Model II. von Karman's Equations

Let us consider the set of equations:

$$u \frac{\partial u}{\partial r} + w \frac{\partial u}{\partial z} - \frac{v^2}{r} = -\frac{1}{\rho}\frac{\partial p}{\partial r} + \nu\left\{\frac{\partial^2 u}{\partial r^2} + \frac{\partial}{\partial r}\left(\frac{u}{r}\right) + \frac{\partial^2 u}{\partial z^2}\right\}$$

$$u \frac{\partial v}{\partial r} + \frac{1}{r} uv + w \frac{\partial v}{\partial z} = \nu\left\{\frac{\partial^2 v}{\partial r^2} + \frac{\partial}{\partial r}\left(\frac{v}{r}\right) + \frac{\partial^2 v}{\partial z^2}\right\}$$

$$u \frac{\partial w}{\partial r} + w \frac{\partial w}{\partial z} = -\frac{1}{\rho}\frac{\partial p}{\partial z} + \nu\left\{\frac{\partial^2 w}{\partial r^2} + \frac{1}{r}\frac{\partial w}{\partial r} + \frac{\partial^2 w}{\partial z^2}\right\}$$

These are momentum equations representing steady, axially-symmetric incompressible viscous flow of a homogeneous fluid. Here

† The values obtained by PIS were checked [9] by an implicit scheme where matrix factorization was used for solution.

(r, θ, z) = cylindrical polar co-ordinate system

u, v, w = velocity components along r, θ, z increasing directions respectively

p = pressure

ρ = density

ν = kinematic coefficient of viscosity.

The equation of continuity is:

$$\frac{\partial}{\partial r}(ru) + \frac{\partial}{\partial z}(rw) = 0.$$

von Karman considered motion of an infinite disk in steady rotation with a constant angular velocity ω and introduced the following transformations:

$$u = \omega r F(\zeta), \quad w = (\nu\omega)^{1/2} H(\zeta), \quad v = \omega r G(\zeta), \quad \rho = -p\nu\omega P(\zeta)$$

where $\zeta = (\omega/\nu)^{1/2} z$. The boundary conditions are: $z = 0$, $u = w = 0$, $v = r\omega$ as $z \to \infty$, $u, v \to 0$. von Karman obtained the following ordinary differential equations:

$$F'' - HF' - F^2 + G^2 = 0, \quad G'' - HG' - 2FG = 0,$$

$$P' - HH' + H'' = 0 \quad \text{and} \quad H' + 2F = 0,$$

which may be further reduced to:

$$H''' - HH'' + 0.5(H')^2 - 2G^2 = 0$$

$$G'' - HG' + H'G = 0.$$

The boundary conditions are:

$$G(0) = 1, \quad H(0) = H'(0) = G(\infty) = H'(\infty) = 0.$$

By introducing $Y = H'$, this system may be reduced to a coupled integro-differential equations and approximated by finite-differences and solved by PIS. The results are compared with those obtained by Benton [2] and given below:

	F		G		H	
η	BENTON	PIS	BENTON	PIS	BENTON	PIS
0.1	0.0462	0.0462165	0.9386	0.938567	-0.0048	-0.004977
0.5	0.1536	0.153592	0.7076	0.70759	-0.919	-0.920607
1.0	0.1802	0.180123	0.4766	0.476651	-0.2655	-0.265421
1.5	0.1559	0.155843	0.3132	0.31324	-0.4357	-0.435795
2.0	0.1189	0.118849	0.2033	0.203367	-0.5732	-0.573111
2.5	0.0848	0.0847674	0.1313	0.131271	-0.6745	-0.67458
3.0	0.0581	0.0581412	0.0845	0.084503	-0.7452	-0.745178
3.5	0.0389	0.0389406	0.0544	0.0543247	-0.7932	-0.793351
4.0	0.0257	0.0257	0.0349	0.0349	-0.8251	-0.825046

CONCLUSION

PIS is applicable only to nonlinear systems. For linear systems, PIS reduces to ordinary Gauss-Seidel iterations, because if we represent a linear system as:

$$x_i = g_i(x_1, \ldots, x_n)$$

the functional g_i will be independent of x_i and hence ω_i^k, the perturbation parameters will be zero identically. Even for a nonlinear system for some i, g_i may be independent of x_i. In such a case the rate of convergence of PIS will be slowed down.

It is quite apparent that the algorithm of PIS is simple and requires no linearization at any stage of its application. Since no Jacobian is computed, no matrix will be stored and consequently computer memory storage requirement is small. Furthermore, since in the vicinity of the root PIS has a quadratic rate of convergence computational time should be small in comparison with most other methods. Recently [10] relaxation parameters were introduced in PIS to accelerate its rate of convergence. This was done for boundary-layer-type flows past bodies with arbitrary shapes. However optimized relaxation parameters to generate the fastest rate of convergence are yet to be found.

In [7], [8], and [9] PIS has been compared with other existing methods. It has been found to be more effective than similar functional iterations.

It is quite clear that PIS has not yet been directly applied to a physiological problem. Most applications have been made to models for which some reasonably valid solutions are available. These results have showed that the algorithm of PIS is quite powerful to solve nonlinear systems effectively. The two nonlinear models presented in the third section have certain characteristic features quite similar to those of physiological flows as discussed in the first section. Thus it may be expected that researchers in mathematical physiology might find it worthwhile to look into the simple algorithm of PIS and apply it to their nonlinear models.

With regard to mathematical modeling of the motion of blood it is quite apparent (Figure 5.1a) that the geometry of motion is very complex. This suggests that if cartesian co-ordinate system is employed in the finite difference solution, the grid-error generated by the net spacings will be large. In such cases body-fitted curvilinear co-ordinate systems as developed by Thompson [31] may be used. The method requires numerical solution of two nonlinear elliptic partial differential equations which could generate co-ordinate lines coincident with all boundaries belonging to a multi-connected region and having arbitrary shapes. These curvilinear co-ordinates are generated in such a way that one co-ordinate is specified to be constant on each of the boundaries, and the distribution of the other is specified along the boundaries. The method can be applied to generate body-fitted curvilinear co-ordinate system around unsteady boundaries such as deforming bodies and free surfaces. It is thoroughly well known that blood vessels are deforming bodies with constrictions and the whole system is time dependent. Also for immersed boundaries such as prosthetic heart valves, we have rigid unsteady bodies in a viscous flow. Thus Thompson's scheme will still be applicable here.

It now appears that a numerical scheme consisting of: (i) Thompson's method of co-ordinate transformation which will generate body-fitted curvilinear co-ordinate system and (ii) the algorithm of PIS for numerical solution of the resulting non-linear model may be a viable numerical procedure to solve problems on physiological flows.

ACKNOWLEDGMENT

This project has been partially supported by a grant from Eastern Illinois University and a fellowship from the Centre of Advanced Studies in Applied Mathematics of the University of Calcutta.

I thank Professor P. K. Ghosh, Head of the Department of Applied Mathematics of Calcutta University, for his many valuable suggestions in my work. I thank Dr. Roland Spaniol, Director of the Computing Center of Eastern Illinois University, without whose help I could never have completed the computational works presented in this article. Thanks are also due to Prof. T. J. Mueller, Professor of Aerospace and Mechanical Engineering for his valuable suggestions on this project.

Finally, I thank my wife Sabita for her great patience and understanding of my studies.

REFERENCES:

1. Ames. W. F., Numerical Methods for Partial Differential Equations, Barnes and Noble Inc., New York, 1969.

2. Benton, E., On the Flow Due to a Rotating Disk, J. Fluid Mechanics, 24: 781-800, part. 4 (1966).

3. Cataloglu, A., Gould, P. L. and Clark, R. E., Validation of a Simplified Mathematical Model for the Stress Analysis of Human Aortic Heart Valves, J. Biomechanics 8: 347-348 (1975).

4. Charm, S. and Kurland, G. S., Viscometry of Human Blood for Shear Rates of 0 - 10^5 sec.$^{-1}$, Nature 206: 617 (1965).

5. Chow, J. C. and Soda, K., Laminar Flow in Tubes with Constriction, Phys. Fluids 15: 1700-1706 (1972).

6. Daly, B. J., A Numerical Study of Pulsatile Flow Through Constricted Arteries, in Proceedings of the Fourth International Conference on Numerical Methods in Fluid Dynamics, Lecture Notes in Physics, Springer-Verlag, New York, June 1974, 117-124.

7. Dey, S. K., Numerical Solution of Nonlinear Implicit Finite Difference Equations of Flow Problems by Perturbed Functional Iterations, in Computational Fluid Dynamics, W. Kollmann, editor, Hemisphere Publishing Corporation, New York, 1980, 543-602.

8. Dey, S. K., Numerical Solution of Nonlinear Equations with Real and Complex Variables, J. Franklin Institute 307: 21-30 (1979).

9. Dey, S. K., Perturbed Iterative Solution of Nonlinear Equations with Applications to Fluid Dynamics, J. Computational and Applied Mathematics 3: 15-30 (1977).

10. Dey, S. K., Numerical Solution of Boundary-Layer-Type Equations by a Perturbed Iterative Scheme, Eastern Illinois University, May 1980.

11. Dey, S. K., Computer Programming of Fluid Flow Phenomena Past a Circular Cylinder, COED Trans. of ASEE VIII: 12, 121-127 (Dec. 1976

12. Fromm, J. E., A Method for Computing Non-Steady Incompressible Viscous Fluid Flows, Report LA-2910, Los Alamos Scientific Lab., Los Alamos, New Mexico (1963).

13. Guyton, A. C., Textbook of Medical Physiology, Saunders, London, 1966.

14. Hopkins, G. R. and Lin, Y. T., Velocity and Pressure Distribution Around a Spherical Particle Transported by a Viscous Fluid Through a Small Tube, Biomechanics Symposium, Georgia Institute of Technology, ASME, 1973, 59-60.

15. Lee, J. S. and Fung, Y. C., Flow in Locally Constricted Tubes at Low Reynolds Numbers, ASME J. Applied Mechanics 9: 9-16 (1970).

16. Lew, H. S., The Use of Entry Flow Equations in Studying the Uniform Entry Flow, J. Biomechanics 6: 205-214 (1973).

17. Lighthill, J., Physiological Fluid Dynamics, International Centre for Mechanical Sciences, Udine, Springer-Verlag, New York, 1971.

18. MacDonald, D. A., On Steady Flow Through Modeled Vascular Stenoses, J. Biomechanics, 12: 13-20 (1979).

19. Merrill, E. W., Rheology of Blood, Physiological Review 49: 874-877 (1969).

20. Meuller, T. J., Application of Numerical Methods to Physiological Flows. Lecture Series 87 on Computational Fluid Dynamics, von Karman Institute for Fluid Dynamics (March 1976). (Also published in Numerical Methods in Fluid Dynamics, Hemisphere Publishing Corp., Washington, D.C., 1978, Chapter 2, 89-154).

21. Meuller, T. J. and Underwood, F. N., Numerical Study of the Steady Axisymmetric Flow Through a Disk-Type Prosthetic Heart Valve in an Aortic-Shaped Chamber, J. Biomedical Engineering 101: 198-204 (1979).

22. Meuller, T. J., Private Communication.

23. Ortega, J. M. and Rheinboldt, W. C., Iterative Solution of Nonlinear Equations in Several Variables, Academic Press, New York, 1970.

24. Patankar, S. V. and Spalding, D. B., Heat and Mass Transfer in Boundary Layers, Intertext Book, London, 1970.

25. Peskin, C. S., Flow Patterns Around Heart Valves: A Numerical Method, J. Computational Physics 10: 252-271 (1972).

26. Roache, P. J., Computational Fluid Dynamics, Hermosa Publishers, Albuquerque, New Mexico, 1975.

27. Rosenhead, L., Laminar Boundary Layers, Clarendon Press, 1963.

28. Rubinow, S. L., Mathematical Problems in the Biological Sciences, Regional Conference Series in Applied Mathematics, SIAM, 1973.

29. Shah, V. L. and Soto, R., Numerical Solution of Blood Flow in the Entrance of a Tube, in Oxygen Transport to Tissue, Duane F. Bruley and Haim J. Bicher, editors, Plenum Publishing Corporation, New York, 867-872.

30. Shah, V. L. and Soto, R., Non-Newtonian Blood Flow in the Entrance Region of a Tube, Computers and Fluids 2: 273-284 (1974).

31. Thompson, J. F., Numerical Solution of Flow Problems Using Body-fitted Co-ordinate Systems, in Computational Fluid Dynamics, W. Kollmann, editor, Hemisphere Publishing Corporation, New York, 1980, 1-98.

32. Thompson, J. F., Numerical Solution of the Incompressible, Two-dimensional, Time-dependent Navier-Stokes Equations for a Body Oscillating in Pitch in a Moving Fluid, Research Note No. 86, Aerospace Eng. Dept., Mississippi State University, Mississippi State, Miss. (Oct. 1968).

33. Thompson, J. F., Thames, F. C., Mastin, C. W. and Shanks, S. P., Use of Numerically Generated Body-Fitted Co-ordinate Systems for Solution of the Navier-Stokes Equation, AIAA 2nd Computational Fluid Dynamics Conference Proceedings, Hartford, Connecticut, June 19-20, 1975, 68-80.

34. Thurston, G. B., The Rheology of Blood in Oscillatory and Steady Flow, Biomechanics Symposium, Georgia Institute of Technology, ASME 1973, 7-8.

35. Yoganathan, A. P., Corcoran, W. H. and Harrison, E. C., In Vitro Velocity Measurements in the Vicinity of Aortic Prostheses, J. Biomechanics 12: 156-164 (1979).

BIFURCATIONS AND THE CONTROL OF DEVELOPMENTAL TRANSITIONS: EVOLUTION OF THE CYCLIC AMP SIGNALLING SYSTEM IN THE SLIME MOLD DICTYOSTELIUM DISCOIDEUM

Albert Goldbeter
Faculté des Sciences, Université Libre de Bruxelles, 1050 Brussels, Belgium (On leave at the Department of Biochemistry, University of California, Berkeley, California)

INTRODUCTION

Since the early work of Turing [44], many studies have been devoted to the analysis of morphogenesis as a process of spatial or spatiotemporal self-organization beyond instability of a nonequilibrium, homogeneous stationary state in a system where chemical reactions are coupled to diffusion [40]. Thus Kauffman et al. [25] have recently suggested that the sequential appearance of compartments in developing imaginal discs of Drosophila originates from a succession of spatial dissipative structures which arise as the size of the discs augments. Another example, studied by Keller and Segel [26], concerns the aggregation of starving Dictyostelium discoideum amoebae: for a critical value of the secretion of chemotactic factor by the cells, the homogeneous distribution of the population becomes unstable and a new, inhomogeneous structure arises in which the amoebae divide in a number of separate aggregation territories and collect around centers.

In these morphogenetic processes, time is always a hidden parameter since the effective parameters which govern the breaking of spatial symmetry--e.g., the size of the system, or the secretion rate of a chemotactic factor--themselves vary as a function of time.

If we now disregard the emergence of spatial patterns and look at the development of biological systems along the temporal dimension only, we note that differentiation consists in the establishment of new functions or modes of dynamic behavior. The question arises as to whether these homogeneous developmental transitions occur smoothly or, in contrast, represent a process of self-organization beyond some critical point of instability. In many instances functional differentiation is brought about by switching on or off a number of genes. This process results in the synthesis of specific proteins in response to changes, which may be continuous, in the concentration of one or several biochemical signals that control gene expression. However, the control of the developmental transitions may also involve bifurcation phenomena analogous to those which produce patterns of spatial organization in morphogenesis.

The role of bifurcations in developmental transitions is well exemplified by the cellular slime mold D. discoideum whose aggregation was mentioned earlier as an example of breaking of spatial symmetry. Aggregation occurs by a chemotactic response to cyclic AMP (cAMP) signals [32]. These signals are emitted autonomously and periodically by certain cells which behave as aggregation centers ([12], [43]). Other cells are capable of relaying the chemotactic signals, i.e., when hit by a cAMP signal they synthetize and subsequently release a pulse of cAMP ([1], [12], [43]). Both relay [42] and oscillations ([13], [16]) of cAMP have also been demonstrated in D. discoideum suspensions. It should be noted that other species of cellular slime molds can aggregate without oscillation or relay of the chemotactic signal; thus in Dictyostelium minutum the propagation of the chemotactic factor by simple diffusion results in the formation of a large number of small aggregation territories [12].

As shown by experiments in fields of amoebae aggregating on agar [9] and in cell suspensions [14], the cAMP signalling system which governs intercellular communication in D. discoideum undergoes a sequence of developmental transitions during the 8h-interphase which separates starvation from aggregation. At the beginning of starvation, the cells are unable to respond in any way to cAMP pulses. Some two hours later they respond chemotactically to these signals. Amoebae become able to relay cAMP pulses after two to four hours. Only after six hours do they acquire the capability to produce autonomous, periodic cAMP signals. Cells lose this capability later on. D. discoideum always follows the above sequence, although the timing of the various events exhibits much variation, depending on the strain and on the experimental conditions.

Here, the sequence of developmental transitions no relay-relay-oscillations is analyzed in an allosteric model for the cAMP signalling system in D. discoideum, proposed by Goldbeter and Segel [22]. This model is presented in the following section. The dynamic behavior of the model is investigated in the third part as a function of two of the most important parameters, namely, adenylate cyclase and phosphodiesterase. These enzymes respectively produce and hydrolyze cAMP. It is proposed that the signalling system undergoes the sequential transitions as it moves on a developmental path in the adenylate cyclase-phosphodiesterase space [23]. The predicted path corresponds to the continuous rise observed after starvation for both enzyme activities.

Two alternative models for the signalling system are briefly examined in the fourth section of the paper. They are based, respectively, on desensitization of the cAMP receptor and on the possible activation of adenylate cyclase by a cAMP-dependent covalent modification. As in the allosteric mechanism, oscillations and relay of cAMP occur in these models when the positive feedback exerted by cAMP on adenylate cyclase is taken into account. The relevance of the above results to the control of developmental transitions is discussed in the last section of the paper.

ALLOSTERIC MODEL FOR THE CYCLIC AMP SIGNALLING SYSTEM IN D. DISCOIDEUM

The model is based on the following observations on the signalling system of *D. discoideum* (see ref. [7], [15], [39] for a review):

(a) Cells possess membrane-bound cAMP receptors which face the extracellular medium;

(b) Upon binding to these receptors cAMP activates adenylate cyclase which is probably bound to the inner face of the plasma membrane (the mechanism of this activation is still unknown) [41];

(c) The amoebae secrete cAMP into the extracellular medium where it is hydrolyzed by the membrane-bound and extracellular forms of a phosphodiesterase.

In the model represented in Fig. 6.1, the receptor-enzyme complex is treated as a membrane-bound allosteric protein with catalytic and regulatory (receptor) sites facing, respectively, the inside volume of the cell and the extracellular medium. For definiteness it is assumed that the oligomeric complex possesses two subunits of both the catalytic and regulatory types, and obeys the concerted model for allosteric enzymes [38]. A similar assumption was made in an earlier model for the oscillatory adenylate cyclase reaction [17]. In conditions corresponding to the experiments with continuously stirred *D. discoideum* suspensions, the time evolution of the system is governed by a set of three ordinary differential equations ([22],

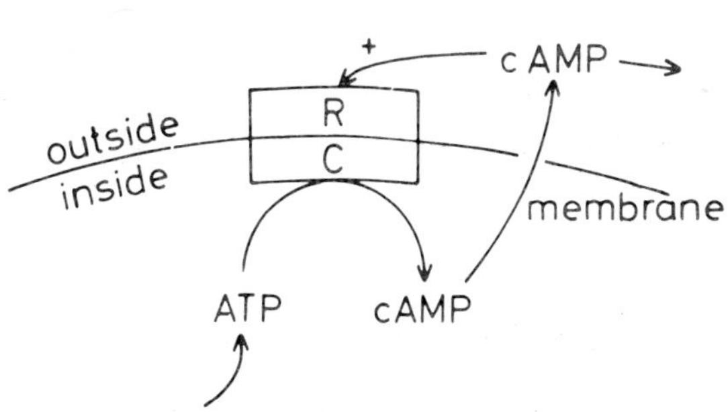

Fig. 6.1. Model for the cAMP signalling system in *D. discoideum* [22]. R and C denote the cAMP receptor and the functionally coupled adenylate cyclase, respectively. Cyclic AMP is synthesized intracellularly from ATP and transported into the extracellular medium where it is hydrolyzed by phosphodiesterase. The + sign refers to the activation of adenylate cyclase upon binding of cAMP to the receptor.

[23]):

$$d\alpha/dt = v - \sigma\phi - k'\alpha$$

$$d\beta/dt = q\sigma\phi - k_t\beta$$

$$d\gamma/dt = (k_t\beta/h) - k\gamma \tag{6.1}$$

with

$$\phi = \alpha(1 + \alpha)(1 + \gamma)^2/[L + (1 + \alpha)^2(1 + \gamma)^2] \tag{6.2}$$

The variables α, β, and γ denote the normalized concentrations of intracellular ATP, intracellular cAMP and extracellular cAMP divided respectively by K_S, K_p, and K_P where K_S and K_P are the Michaelis constant of adenylate cyclase for ATP and the dissociation constant of the cAMP receptor. Also $q = K_S/K_P$, while σ and v are the maximum cyclase activity and a constant ATP input, divided by K_S. The apparent first-order rate constants k_t and k relate to the transport of cAMP across the cell membrane and to the phosphodiesterase reaction, which are both considered as linear processes (both the extracellular and membrane-bound forms of phosphodiesterase are lumped together into parameter k); L is the allosteric constant of adenylate cyclase; h is a dilution factor. The term $-k'\alpha$ in the evolution equation for α represents a modification of the system originally considered in [22]. This term relates to the utilization of ATP in reactions other than that catalyzed by adenylate cyclase. Alternatively, when combined with parameter $v = k'\alpha_0$, it gives the term $k'(\alpha_0 - \alpha)$ which can be viewed as expressing a possible transport of ATP from a constant intracellular pool (α_0) to the site of cAMP synthesis at the membrane.

It should be noted that evidence has been found for a direct activation of adenylate cyclase by cAMP in broken D. discoideum cells [31]. The dynamic behavior of a system based on the activation of adenylate cyclase by intracellular cAMP is similar to that of the system governed by Equations (6.1). Indeed, a mechanism based on the direct activation of the enzyme by its reaction product gives rise to oscillatory and excitable behavior, as shown in a model for the product-activated phosphofructokinase reaction which gives rise to glycolytic oscillations in yeast and muscle ([18], [19]).

DEVELOPMENTAL PATH FOR THE SIGNALLING SYSTEM IN THE ADENYLATE CYCLASE-PHOSPHODIESTERASE SPACE

The model governed by Equations (6.1) is capable of several types of dynamic behavior, depending on the values of its parameters. Two of the main parameters of the model are σ and k, i.e., the activity of adenylate cyclase and the rate constant of phosphodiesterase. It will be convenient to discuss the different types of dynamic behavior in the parameter space formed by σ and k, for a given value of the other parameters.

When the steady-state solutions of Equations (6.1) are subjected to linear stability analysis, the behavioral diagram of Figure 6.2 is obtained in the (σ, k) space [23]. In region A, the system evolves toward a stable steady state corresponding to low levels of both intra- and extracellular cAMP. The signalling system in such a state is unable to amplify cAMP pulses applied externally. In region B, the stable steady state is excitable as the signalling system amplifies in a pulsatory manner supra-threshold perturbations in the extracellular cAMP concentration (Figure 6.3a). This property accounts for the capability of D. discoideum to relay cAMP signals ([1], [12], [42], [43]).

The steady state becomes unstable in domain C; the system then undergoes sustained oscillations in time (Figure 6.3b). This behavior accounts for the fact that centers of aggregation territories as well as aggregation-competent cells in suspensions are capable of producing, in an autonomous manner, periodic pulses of cAMP ([7], [12], [13], [16], [43]). In domain D, the signalling system reaches a stable, non-excitable steady state corresponding to high levels of both intra- and extracellular cAMP.

Finally, in region E, the system admits multiple steady states for a given value of the pair (σ, k). Depending on the values of these parameters one or two of the three steady states are stable [18]. In the former instance, the system in the stable state is excitable and its behavior is analogous to that depicted in Figure 6.3a. When two steady states are stable, the system can switch between them upon perturbation of the extracellular cAMP concentration. The latter phenomenon corresponds to the transition from a low to a high steady level of both intra- and extracellular cAMP for a given maximum activity of adenylate cyclase and phosphodiesterase. Although further experiments are needed to investigate the possibility of such all-or-none transitions, the phenomenon might be related to the observation [24] that adenylate cyclase can undergo a transition from an "inactive" to an "active" form upon stimulation by cAMP pulses in D. discoideum suspensions.

The partitioned parameter space (σ, k) suggests a developmental path for the signalling system of D. discoideum. This path accounts for the behavioral sequence no relay-relay-oscillations. The trajectory followed in the parameter space (see arrow in Figure 6.2) relies on the notion that continuous changes in the activity of adenylate cyclase and phosphodiesterase after starvation may bring about discontinuous transitions in the dynamic properties of the signalling system.

Initially, both adenylate cyclase and phosphodiesterase possess a low activity. Let us suppose that cells are then located in region A in the diagram of Figure 6.2. Amoebae in this state are unable to relay cAMP pulses. An increase in adenylate

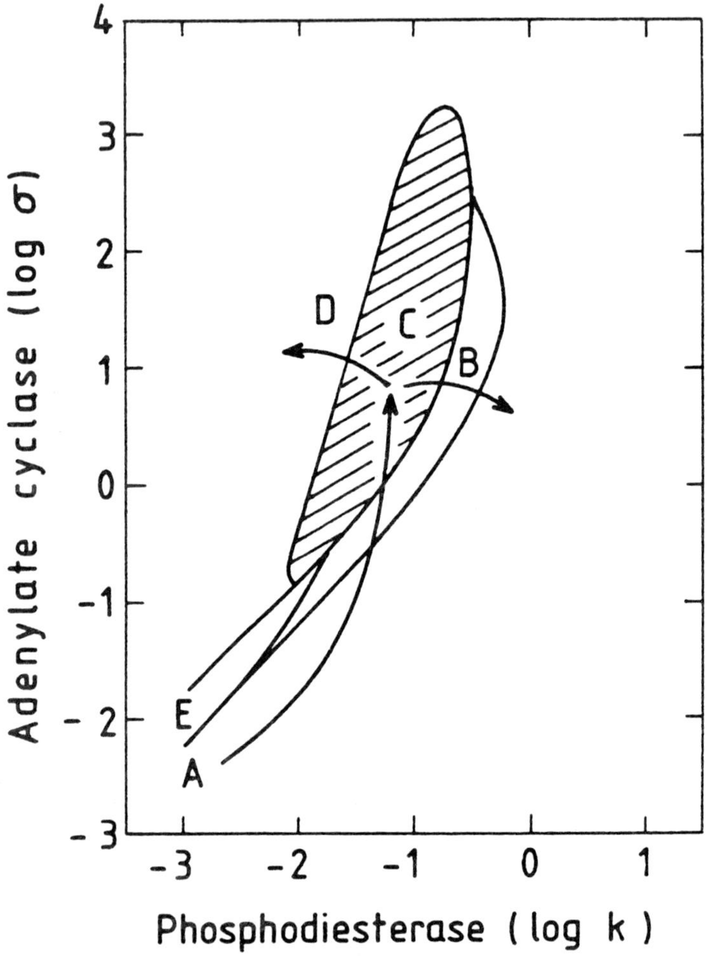

Fig. 6.2. The partitioned adenylate cyclase-phosphodiesterase activity plane (from ref. [23]). If a parameter (σ, k) is located in domain A, the corresponding behavior of the model governed by Equations (6.1) is steady low cAMP secretion. Domains B, C, and D correspond respectively to relay, autonomous oscillation, and high steady secretion of cAMP. Multiple steady states are found in domain E. During the course of starvation the parameter state of the system is represented by a point which moves slowly along the developmental path (arrow). The path crosses successively the no-relay, relay and oscillatory domains; upon ceasing to oscillate the system may either return to the excitability domain or settle to a high steady cAMP secretion. Oscillations in domain C occur around a nonequilibrium unstable steady state [22]. The stability properties of the steady state were determined by linear stability analysis, whereas the relay domain was constructed by numerical simulations of System (6.1) as described in ([18], [22]). The cAMP signal was simulated by setting the initial concentration of extracellular cAMP as ten times the dissociation constant K_p of the cAMP receptor, i.e., $\gamma = 10$; relay occurred when the system responded by producing a peak in γ exceeding this value. The diagram was established for the parameter values $v = 0.04\ s^{-1}$, $k_t = 0.4\ s^{-1}$, $k' = 10^{-3}\ s^{-1}$, $h = 10$, $L = 10^6$, $K_p = 10^{-7}M$, $q = 100$ ([22], [23]).

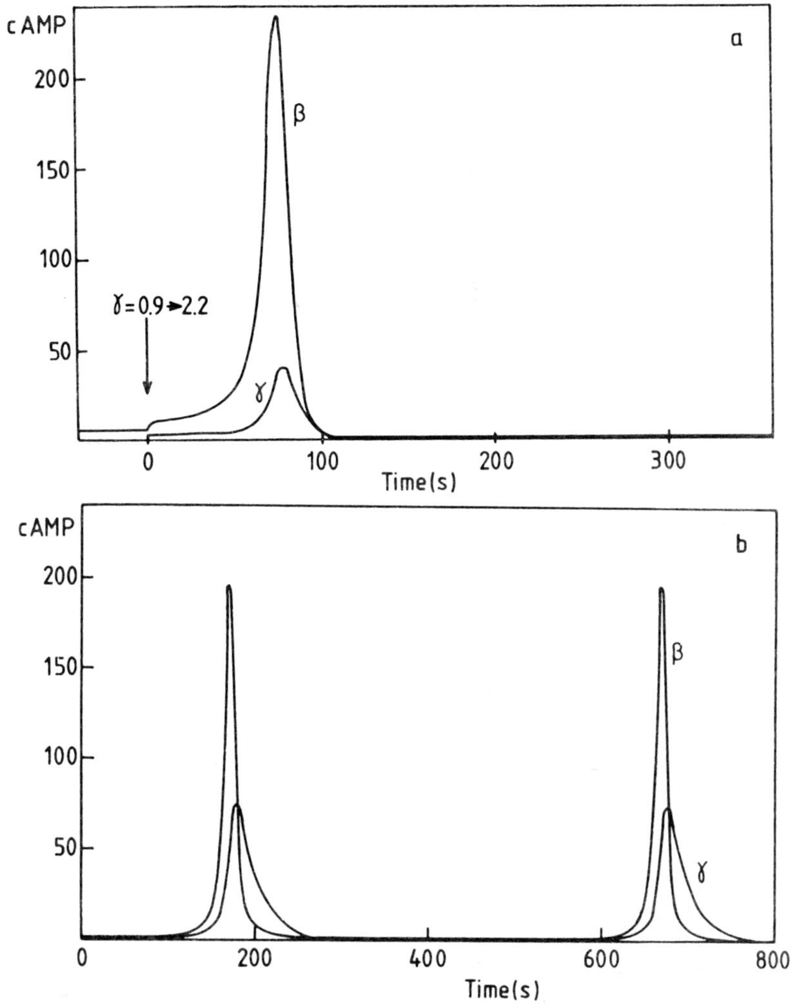

Fig. 6.3. Relay of a cAMP signal upon addition of a suprathreshold cAMP pulse at time zero (a), and sustained oscillations (b); β and γ denote the normalized concentrations of intracellular and extracellular cAMP. The curves, obtained by integration of Equations 6.1, illustrate the behavior of the signalling system in domains B and C of Figure 6.2 for $\sigma = 10\ s^{-1}$, $k = 0.2\ s^{-1}$ and $\sigma = 10\ s^{-1}$, $k = 0.05\ s^{-1}$, respectively.

cyclase and phosphodiesterase activity will then bring the cells into the relay domain B. The area of domain B increases with the magnitude of the cAMP signal; the boundary of the relay domain is thus somewhat fuzzy, in compliance with the fact that it does not correspond to a bifurcation since the steady state, though excitable, remains stable.

In contrast, the entrance of the signalling system into domain C upon further increase in σ and k truly corresponds to the bifurcation of a self-sustained periodic regime. Physiologically acceptable values of the parameters can be found for which the period of the oscillations is of the order of minutes [22] (see Figure 6.3b).

The diagram in Figure 6.2 allows one to discuss the nature of the cells that become the centers of aggregation. In terms of the model, these cells are the first to enter the instability domain of the adenylate cyclase reaction--i.e., domain C--and thereby begin to synthesize and release cAMP periodically ([17], [22]). It has been observed that when centers are removed from aggregation territories they can be replaced in that role by some relaying cells. In the model, accordingly, the rate of increase in adenylate cyclase and phosphodiesterase activity is likely to be distributed among the population of starved cells. Upon removal of the centers, the two enzyme activities will reach the critical value required for instability in other cells which will then enter the oscillatory domain and will become new aggregation centers.

Experiments indicate that after several hours cells in aggregation territories lose the capability of becoming centers of aggregation. This result is confirmed by studies in suspensions which show that D. discoideum amoebae cease to oscillate after several hours and thereupon become excitable again [14]. In the diagram of Figure 6.2, such an evolution would correspond to a return from region C into the relay domain B. The biochemical correlate of such a move would be an increase in the activity of phosphodiesterase. Such an increase could correspond to the second wave of phosphodiesterase synthesis which has previously been reported [30].

Upon quitting the oscillatory domain, cells could also move towards domain D provided that the activity of phosphodiesterase would drop to a sufficiently low level. Cells in domain D are characterized by a high steady secretion of cAMP. As indicated above, such a state is not reached in the experiments with cell suspensions. Previous evidence indicated that cells forming the tips of late aggregates behave as constant high sources of cAMP [10]. Recent work suggests, however, that cells at the slug stage are still excitable since relay of chemotactic signals seems to play a role even at that phase of the morphogenetic program ([2], [11]).

The above results suggest that the sequential transitions observed during development of the cAMP signalling system may originate from a continuous change in the activity of two key enzymes, namely, adenylate cyclase and phosphodiesterase. In support of such a view is the fact that an increase in the activity of the two enzymes represents one of the first biochemical events triggered by starvation [34]. Adenylate cyclase undergoes a 40-fold increase in activity some four to five hours after the beginning of interphase [27]; this time course is paralleled by that of phosphodiesterase ([15], [30], [34]). This increase in enzyme activities fits well with the fact that the developmental path in Figure 6.2 crosses successively the regions of no relay, relay, and oscillations. The subsequent exit from the oscillatory domain to either regions B or D should be correlated with an increase or a decrease in phosphodiesterase activity which might depend on cellular conditions.

Although the evolution of the cAMP signalling system has been discussed with respect to only two parameters, it is clear that it takes place in a multidimensional parameter space. Some fifty genes are required for the completion of aggregation in D. discoideum ([4], [45]). An unknown fraction of these genes governs the mechanism of intercellular communication, but the number of gene products controlling the evolution of the signalling system is likely to exceed two. Adenylate cyclase and phosphodiesterase constitute, nevertheless, the main elements of this system since they relate directly to the production and removal of the signal.

ALTERNATIVE MODELS FOR THE CYCLIC AMP SIGNALLING SYSTEM

In the model analyzed in the above, it is assumed that adenylate cyclase and the cAMP receptor behave as the catalytic and regulatory subunits of an oligomeric protein which obeys the concerted model for allosteric enzymes. Evidence for an allosteric nature of either adenylate cyclase or the cAMP receptor in D. discoideum is still lacking, due to the difficulty of isolating these proteins. The question arises as to whether the allosteric assumption for the enzyme and the receptor is needed for relay and oscillation of cAMP. To answer this question, two alternative mechanisms based on the regulation of a monomeric adenylate cyclase have recently been studied ([21], [35], [36]).

Activation of adenylate cyclase by a cAMP-dependent protein kinase

One way in which the positive feedback exerted on adenylate cyclase by its product could occur involves the phosphorylation of the enzyme by a cAMP-dependent protein kinase (Figure 6.4). The analysis of such a model [35] shows the possibility of relay and oscillations of cAMP (see Figure 6.5) when the nonlinear activation of the protein kinase by cAMP [33] is taken into account. Similar results are obtained when adenylate cyclase is activated by phosphorylation through a cyclic GMP-dependent protein kinase. Indeed, the positive feedback loop would still be present, given that extracellular cAMP elicits the synthesis of a pulse of cyclic GMP upon chemotactic stimulation of D. discoideum [37].

These results remain theoretical since no evidence has yet been found for any covalent modification of adenylate cyclase in D. discoideum.

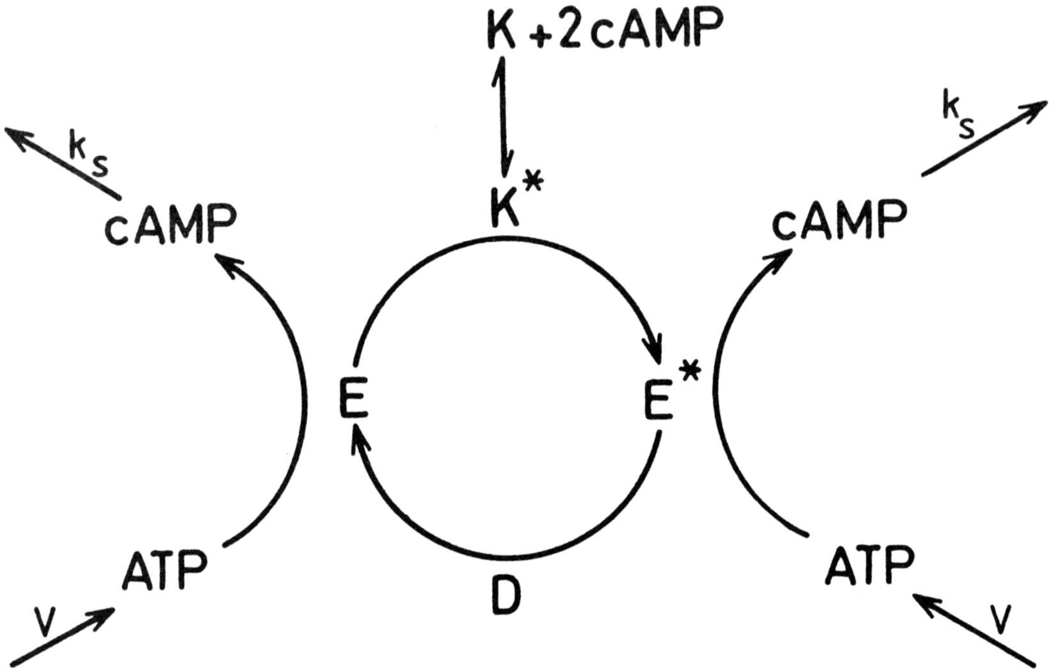

Fig. 6.4. Autocatalytic regulation of adenylate cyclase by covalent modification, giving rise to cAMP oscillations and excitability (from ref. [35]). E and E* are, respectively, the less active and more active forms of adenylate cyclase; K and K* are the inactive and active forms of the protein kinase which phosphorylates E; enzyme D is a phosphatase which reverts this modification. Autocatalysis results from the nonlinear activation of the protein kinase by cAMP. The operation of such a scheme in *D. discoideum* remains hypothetical.

Periodic desensitization of the cAMP receptor

An alternative mechanism for cAMP relay and oscillation is based on the observation [28] that extracellular cAMP induces desensitization of the cAMP receptor. Evidence for the role of such a process in cAMP oscillations has previously been reported in [29]. In the mechanism considered ([21], [36]), the receptor exists in two conformations, active (R) or desensitized (D). Extracellular cAMP binds to the D state with a higher affinity than to the R state, but the latter predominates in the absence of signal. When only the active receptor is considered to couple to adenylate cyclase, thereby activating cAMP synthesis, it is possible to show that oscillations and relay can occur provided that binding of cAMP to the receptor is cooperative.

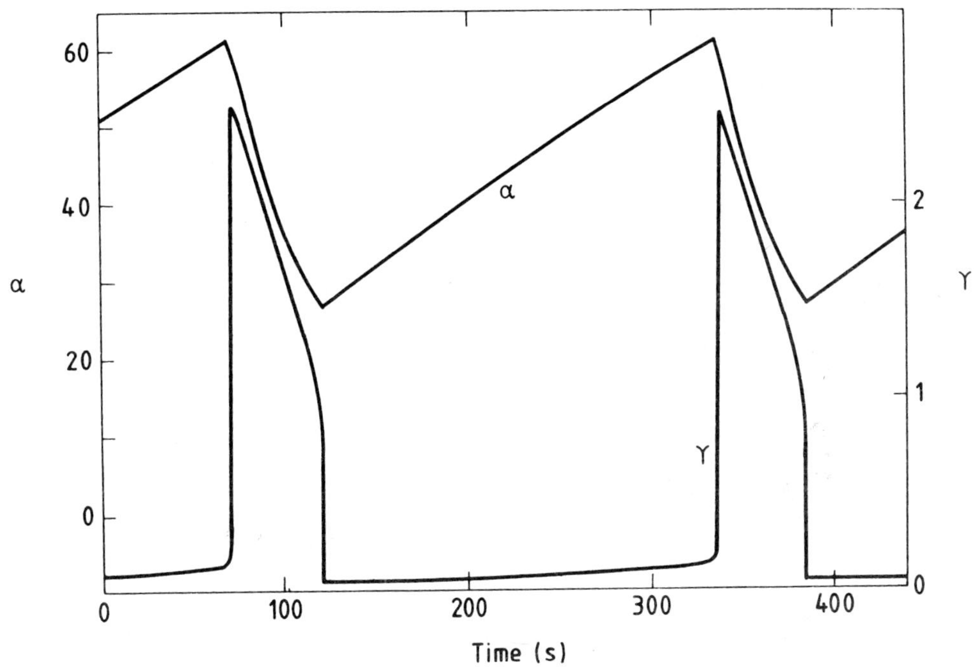

Fig. 6.5. Sustained oscillations of ATP (α) and cAMP (γ) due to a putative phosphorylation of adenylate cyclase by a cAMP-dependent protein kinase (from ref. 35). The model generating these curves is that of Fig. 6.4.

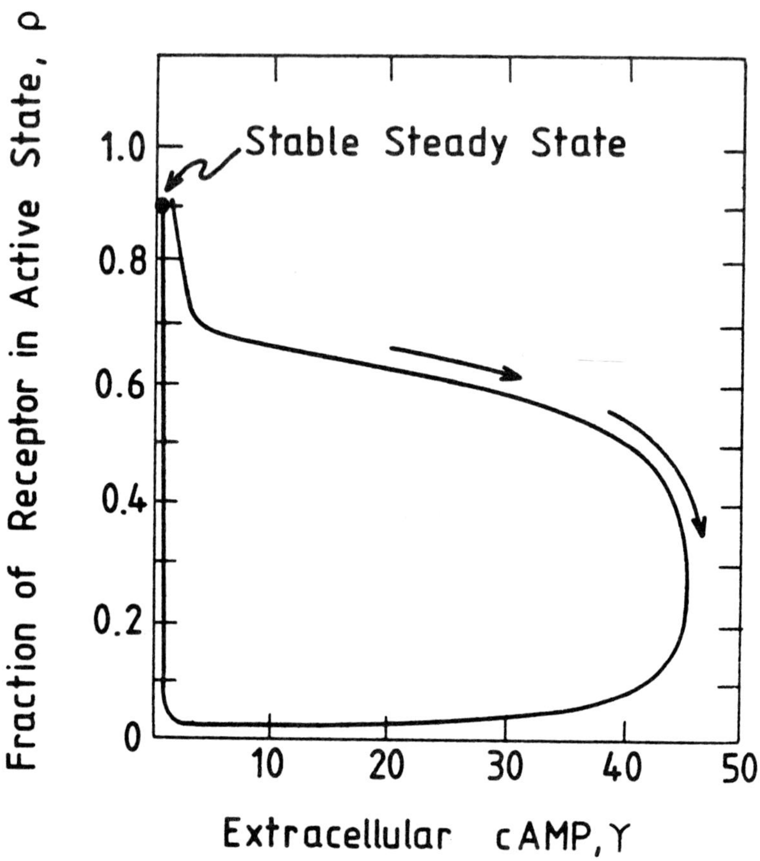

Fig. 6.6. Relay trajectory in the phase plane of the model for desensitization of the cAMP receptor coupled to adenylate cyclase. The horizontal displacement of γ away from the stable steady state corresponds to the addition of a cAMP pulse; in response to such a signal the system synthesizes a pulse of cAMP. The decrease in synthesis is brought about by a drop in the fraction of receptor in the active conformation ([21], [36]).

An interesting feature of this model is that the decrease in cAMP synthesis during oscillations or relay (see Figure 6.6) is brought about not by a decrease in substrate--ATP is held constant here--but by the passage of the active receptor into its desensitized state. The return of the receptor to the active conformation, upon reversal of the desensitization, follows the decrease in extracellular cAMP. In the course of oscillations, the receptor is subjected to periodic desensitization which could account for the periodicity of cAMP binding observed in the experiments [29] (it should be noted that oscillatory binding also occurs in the previous models as a result of cAMP oscillations).

The desensitization model is currently being investigated as to its behavior when the extracellular cAMP level is maintained constant. Recent experiments indicate that cAMP synthesis augments transiently and adapts in response to such "clamped" stimuli; a further increase in cAMP level is needed to elicit a new round of cAMP synthesis [8]. Although these experiments do not correspond to physiological conditions since they suppress the hydrolysis of the signal by phosphodiesterase, they throw light on the signalling mechanism and should be accounted for by any model for the coupling between the cAMP receptor and adenylate cyclase.

Excitable and oscillatory behavior are properties of the three models discussed above. In each model it is the positive feedback exerted on adenylate cyclase by its product cAMP which is the primary cause of the instability that leads to oscillations. It is of course possible that several of the above modes of interaction between adenylate cyclase and the cAMP receptor are simultaneously at work to produce relay and oscillations [22]. The latter two phenomena are always linked; phase plane analysis shows indeed that stable steady states are excitable for some parameter values close to those that produce periodic behavior ([18], [20]).

CONCLUDING REMARKS

During the interphase which separates starvation from aggregation, the cAMP signalling system in D. discoideum successively acquires the capability of relay and autonomous oscillation. The analysis of an allosteric model for the signalling system shows that the continuous variation of adenylate cyclase and phosphodiesterase after starvation can account for the sequential transitions no relay-relay-oscillation.

The developmental path suggested in the adenylate cyclase-phosphodiesterase space obviates the need to invoke the switching of different sets of genes to produce the cAMP relay and oscillatory mechanisms. A continuous increase in the products of a given set of genes is all that is needed to trigger the discontinuous passage between several behavioral domains, one of which corresponds to the bifurcation of self-sustained oscillations.

The predictions made on the basis of the behavioral diagram of Figure 6.2 could be checked experimentally in several ways, primarily by the study of mutants defective in the signalling system. Thus a mutant lacking phosphodiesterase enters the oscillatory domain immediately upon addition of this enzyme [6], as predicted by the model when such cells are initially located in domain D (see Figure 6.2). Another mutant, Fr17, aggregates more rapidly than the wild type due to the early occurrence of cAMP oscillations (in addition to controlling chemotaxis, periodic cAMP pulses control differentiation in interphase amoebae ([7], [14])). This behavior of Fr17 has been linked to a precocious rise in adenylate cyclase activity [5] which could correspond to a faster movement on the developmental path in Figure 6.2. Conversely, the mutant 91A is characterized by a slow rate of emergence of autonomous centers and by a tendency to remain in the relaying mode [10]. The model predicts that the activity of adenylate cyclase is smaller in 91A than in the parent wild type.

The diagram of Figure 6.2 also suggests that the signalling properties of wild type cells in suspensions could be altered by addition of phosphodiesterase or of an inhibitor of this enzyme [23]. It should be possible in this way to switch from relay to oscillations and vice versa, or to suppress both types of phenomena.

Though the developmental path approach has been illustrated by means of a specific model, it is clear that similar results can be established in any model for the signalling system that exhibits relay and oscillation of cAMP. This applies to the two alternative mechanisms considered in this paper, and to a model for the adenylate cyclase reaction in which the enzyme is subjected to another mode of regulation [3]. In each case, the predicted developmental path should be compared with the experimental time course of the parameters governing the transitions. Thus, the number of cAMP binding sites which varies during interphase ([7], [15], [39]) would become one of the key parameters, together with adenylate cyclase and phosphodiesterase, in the model based on receptor desensitization. It will be interesting to determine how the behavioral diagram in the (σ, k) plane is affected in this model by a variation in the receptor concentration.

Several models have by now been proposed to illustrate the existence of self-organization phenomena in developing systems, particularly during morphogenesis. The evolution of the cAMP signalling system of D. discoideum through several behavioral domains provides a further example of the role of bifurcations in development. A particular feature of the slime mold signalling system, revealed by the present analysis, is that some of the parameters governing the developmental transitions have tentatively been identified at the molecular level.

ACKNOWLEDGMENTS

I wish to thank Th. Erneux, J. L. Martiel and L. A. Segel for their collaboration in various parts of this work and for fruitful discussions. This study was supported by a contract "Action de Recherche Concertée" from the Belgian Government.

REFERENCES:

1. Alcantara, F. and Monk, M., Signal propagation during aggregation in the slime mold Dictyostelium discoideum, J. Gen. Microbiol. 85: 321 (1974).

2. Clark, R. C. and Steck, T. L., Morphogenesis in Dictyostelium: an orbital hypothesis, Science 204: 1163 (1979).

3. Cohen, M. S., The cyclic AMP control system in the development of Dictyostelium discoideum, J. Theor. Biol. 69: 57 (1977).

4. Coukell, M. B., Parasexual genetic analysis of aggregation-deficient mutants of Dictyostelium discoideum, Molec. Gen. Genet. 142: 119 (1975).

5. Coukell, M. B. and Chan, F. K., The precocious appearance and activation of an adenylate cyclase in a rapid developing mutant of Dictyostelium discoideum, FEBS Lett. 110: 39 (1980).

6. Darmon, M., Barra, J. and Brachet, P., The role of phosphodiesterase in aggregation of Dictyostelium discoideum, J. Cell Sci. 31: 233 (1978).

7. Darmon, M. and Brachet, P., Chemotaxis and differentiation during the aggregation of Dictyostelium discoideum amoebae, in Taxis and Behavior, Receptors and Recognition (G. Hazelbauer, editor), Series B, Vol. 5: 103, Chapman and Hall, London, 1978.

8. Devreotes, P. N. and Steck, T. L., Cyclic 3', 5' AMP relay in Dictyostelium discoideum. Requirements for initiation and termination of the response, J. Cell Biol. 80: 300 (1979).

9. Robertson, A., Drage, D. J. and Cohen, M. H., Control of aggregation in Dictyostelium discoideum by an external periodic pulse of cyclic adenosine monophosphate, Science 175: 333 (1972).

10. Durston, A. J., Pacemaker mutants of Dictyostelium discoideum, Devel. Biol. 38: 308 (1974).

11. Durston, A. J. and Vork, F., A cinematographical study of the development of vitally stained D. discoideum, J. Cell Sci. 36: 261 (1979).

12. Gerisch, G., Cell aggregation and differentiation in Dictyostelium, Curr. Top. Devel. Biol. 3: 157 (1968).

13. Gerisch, G. and Hess, B., Cyclic-AMP controlled oscillations in suspended Dictyostelium cells: Their relation to morphogenetic cell interactions, Proc. Nat. Acad. Sci. US 71: 2118 (1974).

14. Gerisch, G., Maeda, Y., Malchow, D., Roos, W., Wick, U. and Wurster, B., Cyclic AMP signals and the control of cell aggregation in Dictyostelium discoideum, in Development and Differentiation in the Cellular Slime Molds (P. Cappuccinelli and J. M. Ashworth, editors), p. 105, Elsevier/North Holland Biomed. Press, Amsterdam, 1977.

15. Gerisch, G. and Malchow, D., Cyclic AMP receptors and the control of cell aggregation in Dictyostelium, Adv. Cyclic Nucleot. Res. 7: 49 (1976).

16. Gerisch, G. and Wick, U., Intracellular oscillations and release of cyclic AMP from Dictyostelium cells, Biochem, Biophys. Res. Commun. 65: 364 (1975).

17. Goldbeter, A., Mechanism for oscillatory synthesis of cyclic AMP in Dictyostelium discoideum, Nature 253: 540 (1975).

18. Goldbeter, A., Models for oscillations and excitability in biochemical systems, in Mathematical Models in Molecular and Cellular Biology (L. A. Segel, editor), p. 248, Cambridge Univ. Press, Cambridge, 1980.

19. Goldbeter, A. and Caplan, S. R., Oscillatory enzymes, Ann. Rev. Biophys. Bioeng. 5: 449 (1976).

20. Goldbeter, A., Erneux, T. and Segel, L. A., Excitability in the adenylate cyclase reaction in Dictyostelium discoideum, FEBS Lett. 89: 237 (1978).

21. Goldbeter, A. and Martiel, J. L., Role of receptor desensitization in the mechanism of cAMP oscillations in Dictyostelium, Abstract #1046, Fed. Proc. 39: 1804 (1980).

22. Goldbeter, A. and Segel, L. A., Unified mechanism for relay and oscillation of cyclic AMP in Dictyostelium discoideum, Proc. Nat. Acad. Sci. US 74: 1543 (1977).

23. Goldbeter, A. and Segel, L. A., Control of developmental transitions in the cyclic AMP signalling system of Dictyostelium discoideum, Differentiation, in press, 17 (1980).

24. Juliani, M. H. and Klein, C., A biochemical study of the effects of cAMP pulses on aggregateless mutants of Dictyostelium discoideum, Devel. Biol. 62: 162 (1978).

25. Kauffman, S., Shymko, R. M. and Trabert, K., Control of sequential compartment formation in Drosophila, Science 199: 259 (1978).

26. Keller, E. F. and Segel, L. A., Initiation of slime mold aggregation viewed as an instability, J. Theor. Biol. 26: 399 (1970).

27. Klein, C., Adenylate cyclase activity in Dictyostelium discoideum amoebae and its changes during differentiation, FEBS Lett. 68: 125 (1976).

28. Klein, C., A slowly dissociating form of the cell surface cyclic adenosine 3':5' monophosphate receptor of Dictyostelium discoideum, J. Biol. Chem. 254: 12573 (1979).

29. Klein, C., Brachet, P. and Darmon, M., Periodic changes in adenylate cyclase and cAMP receptors in Dictyostelium discoideum, FEBS Lett. 76: 145 (1977).

30. Klein, C. and Darmon, M., The relationship of phosphodiesterase to the developmental cycle of Dictyostelium discoideum, Biochem. Biophys. Res. Commun. 67: 440 (1975).

31. Klein, C. and Darmon, M., A cAMP-sensitive adenylate cyclase in Dictyostelium discoideum extracts, FEMS Microbiol. Lett. 5: 1 (1979).

32. Konijn, T. M., van de Meene, J. G. C., Bonner, J. T. and Barkley, D. S., The acrasin activity of adenosine-3',5'-cyclic phosphate, Proc. Nat. Acad. Sci. US 58: 1152 (1967).

33. Krebs, E. G. and Beavo, J. A., Phosphorylation-dephosphorylation of enzymes, Annu. Rev. Biochem. 48: 923 (1979).

34. Loomis, W. F., Biochemistry of aggregation in Dictyostelium. A review, Devel. Biol. 70: 1 (1979).

35. Martiel, J. L. and Goldbeter, A., Metabolic oscillations in biochemical systems controlled by covalent enzyme modification, Biochimie, in press (1980).

36. Martiel, J. L. and Goldbeter, A., manuscript in preparation.

37. Mato, J. M. and Malchow, D., Guanylate cyclase activation in response to chemotactic stimulation in Dictyostelium discoideum, FEBS Lett. 90: 119 (1978).

38. Monod, J., Wyman, J. and Changeux, J. P., On the nature of allosteric transitions: A plausible model, J. Mol. Biol. 12: 88 (1965).

39. Newell, P. C., Aggregation and cell surface receptors in cellular slime molds, in Microbial Interactions, Receptors and Recognition (J. L. Reissig, editor), Series B, Vol. 3, p. 3, Chapman and Hall, London, 1977.

40. Nicolis, G. and Prigogine, I., Self-Organization in Nonequilibrium Systems, Wiley-Interscience, New York, 1977.

41. Roos, W. and Gerisch, G., Receptor-mediated adenylate cyclase activation in Dictyostelium discoideum, FEBS Lett. 68: 170 (1976).

42. Roos, W., Nanjundiah, V., Malchow, D. and Gerisch, G., Amplification of cyclic AMP signals in aggregating cells of Dictyostelium discoideum, FEBS Lett. 53: 139 (1975).

43. Shaffer, B. M., The acrasina, Adv. Morphogen. 2: 109 (1962).

44. Turing, A. M., The chemical basis of morphogenesis, Phil. Trans. R. Soc. London Ser. B 237: 37 (1952).

45. Williams, K. L. and Newell, P. C., A genetic study of aggregation in the cellular slime mould Dictyostelium discoideum using complementation analysis, Genetics 82: 287 (1976).

TOWARDS THE ELUCIDATION OF BLOOD COAGULATION

George Karreman†
Department of Physiology, School of Medicine, University of Pennsylvania, Philadelphia, Pennsylvania

ABSTRACT

Blood coagulation is a very complex regulatory process due to the cooperative interactions of many factors and cofactors. The blood coagulation phenomenon is a cascade system consisting of a hierarchy of subsystems which interact non-linearly due to positive feedbacks, resulting from those cooperative processes, and formation of complexes. The results of the mathematical modeling of two important control subsystems of the blood coagulatory system are shown to be in good agreement with experimental data. The significance of recent progress in the theory of critical phenomena and in synergetics for further progress in such complicated open dissipative systems as occur in blood coagulation is pointed out. The importance of electronic biology and cooperative phenomena for fundamental biological processes is stressed. The general significance of cascade systems in science is illustrated with examples for energy production from nuclear physics and biology.

INTRODUCTION

Wound healing is one of the most beautiful examples of regulatory processes in biology and medicine (A. Szent-Györgyi, [50]). The processes of blood coagulation which form an important part of wound healing consist of several cascade systems. Some of the recent models of these cascade systems are shown in Figure 7.1 through 7.4. It is clear that the overall or total blood coagulation system consists of hierarchies of cascade systems with complex, usually nonlinear, interactions. Several of the cascade systems consist of cooperative mechanisms with successive great amplification factors. The occurrence of the latter is not surprising as the positive feedback mechanisms are often the essential components of many fundamental biological processes such as physiological excitation, muscle contraction, morphogenesis and carcinogenesis (G. N. Ling, [25]). Of course many biological states are stabilized by negative feedback mechanisms leading to homeostasis. Biological organisms are open systems often quite far from equilibrium (A. Pacault and C. Vidal, [36]), especially due to the biological processes occurring in them.

---

† This work was supported in part by ONR Contract N00014-75-C-0687.

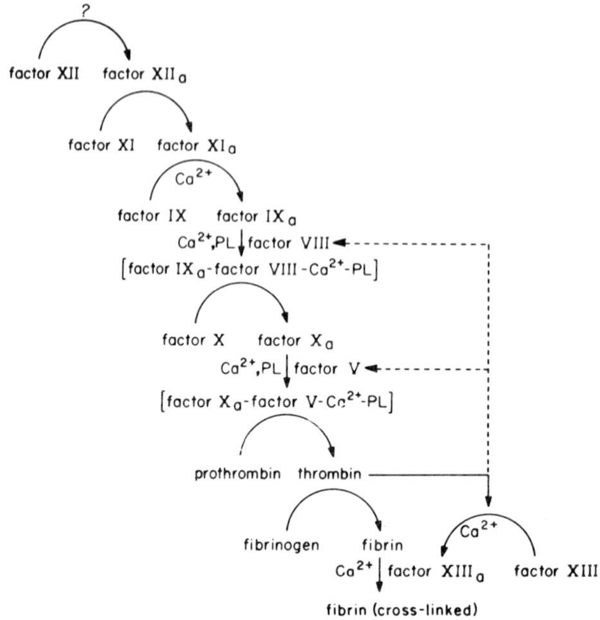

Fig. 7.1. The blood coagulation cascade system from Molecular Mechanisms in Blood Coagulation by E. W. Davie and E. P. Kirby in Current Topics in Cellular Regulation, Vol. 7, p. 52 (B. L. Horecker and E. R. Stadtman, eds.), Acad. Press, New York, 1973. Reprinted by permission.

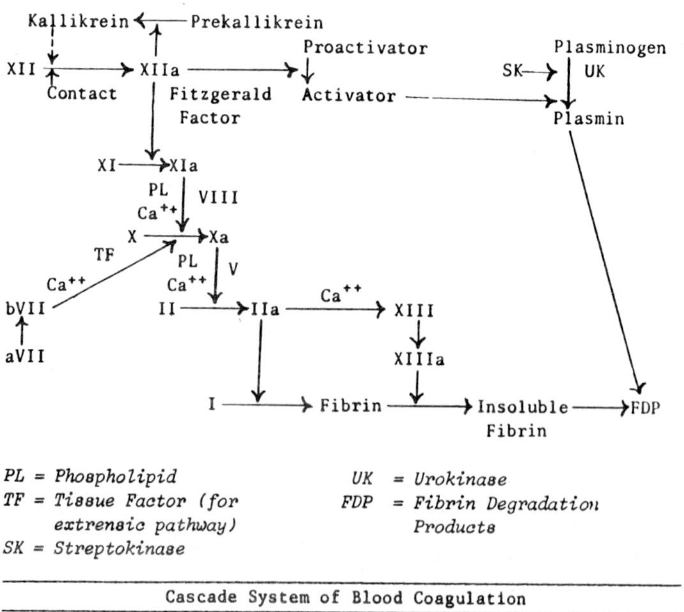

Fig. 7.2. The cascade system of blood coagulation from page 123 of Overview of Blood, C. Bishop, editor, Blood Information Service, Buffalo, N.Y.

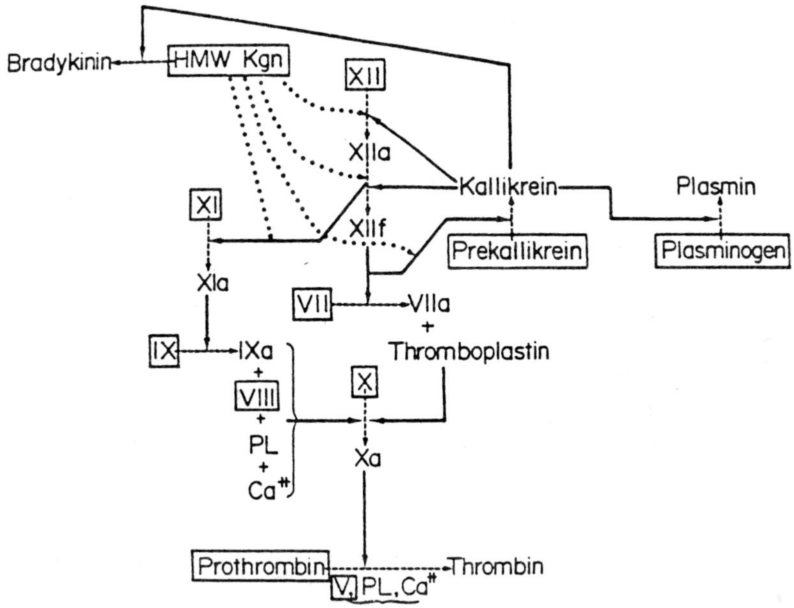

Fig. 7.3. Schematic drawing of the blood coagulation system according to R. W. Colman with whose permission this is redrawn.

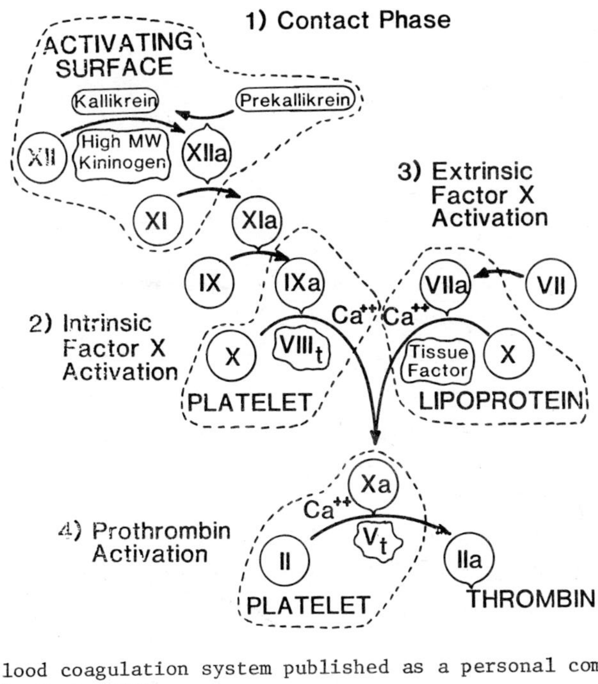

Fig. 7.4. The blood coagulation system published as a personal communication from E. P. Kirby and D. C. B. Mills.

Hence, biological organisms often lead to dissipative structures (P. Glansdorf and I. Prigogine, [4]; G. Nicolis and I. Prigogine, [35]) with changes in temporal and spatial order (H. Haken, [8]). The processes of pattern formation and recognition (H. Haken, [9]) are, of course, also fundamental in biology as for morphogenesis and sensory discrimination. These problems of the production of order in those complex biological systems are intimately related to the change of stability of singularities and limit cycles already studied by H. Poincaré [37] as well as developed into bifurcation theory by A. Adronow et al. ([1]) in mathematics and in phase transitions by L. Landau and V. L. Ginsburg (H. Haken, [8]) for superconductivity. The theory of fluctuations, (L. D. Landahl and E. M. Lifschitz, [24]), especially based on stochastic approaches and the theory of the masterequation (G. Nicolis and I. Prigogine, [35]; H. Haken, [8]) is very important for the solution of these problems. The stochastic fluctuations in these highly non-linear systems play an important role as they may lead to macroscopic changes from a (meta)stable state, which may become unstable, to other (meta)stable ones yielding new ordered structures. For the classification of static instabilities the catastrophe theory of R. Thom ([51]), which originated in theoretical biology (C. L. Waddington, [52], has also been shown to be very important (A. E. R. Woodcock and T. Poston, [56]). This theory, concerned with structural stability, has made not only important contributions to biology, especially embryology by studying bifurcation catastrophes which can be described as singularities of certain potential functions, but also to many other sciences (E. C. Zeeman, [57]). Geometric, topological or qualitative mathematical approaches have become essential for the further development of mathematical biology, as already realized decades ago by N. Rashevsky ([41]) with more recent contributions of R. Rosen ([42], [43], [44]). Important contributions to carcinogenesis, using bifurcation theory and topological approaches, have been made by O. Gurel ([6], [7]).

Of course, much older than these mathematical treatments of biological phenomena have been their interpretations in terms of molecular biology. Though it has often been possible to find explanations of biological phenomena in terms of molecular biology, the most fundamental level in biology is probably submolecular, namely electronic. This view, first suggested by A. Szent-Györgyi ([48], [49]), has been found to be very fruitful as shown by such contributions as A. and B. Pullman ([38], [39]), G. Karreman ([11]-[14]), J. J. Ladik ([21], [22], [23]), G. Del Re and J. J. Ladik ([2]) and many recent papers in the Quantum Biological Supplements of the International Journal of Quantum Chemistry as well as related journals. The difficulties in the rigorous development of quantum chemistry and quantum biology† are

† Quantum biology is at present renamed electronic biology (A. Szent-Györgi, [50]) which may be more appropriate because of a recent evaluation of the present status of quantum theory by R. Rosen [45].

largely of a mathematical nature, especially for the many electron systems occurring in biology though even systems with a few electrons yield also still great mathematical as well as important conceptual difficulties. As for the mathematical difficulties the great progress recently made in the theory of cooperative and critical phenomena may be very useful.

RESULTS AND DISCUSSION

It would be premature to try to model mathematically the whole blood coagulation system at present. This is clear from the four descriptive models given above. As those models do not agree in all respects it is not yet possible to uniquely characterize mathematically the blood coagulation system. After the mathematical characterization of a phenomenon by a model, the latter has to be mathematically identified by determining the best values of the parameters of the model such that the agreement of the responses, which the mathematical model gives due to certain stimuli, with the experimental data is maximized (for instance in a certain least square sense). Unfortunately, neither the mathematical characterization nor the quantitative identification of the whole blood coagulation system is at present yet feasible. Hence, the results of the mathematical modeling of only two important control subsystems of the blood coagulation cascade system will be discussed. Both of these two subsystems control the proteolytic conversion of zymogens under the influence of protein cofactors.

First, a mathematical model is proposed for the activation of prothrombin P by the enzyme factor X_a which includes the feedback mechanism of thrombin T to alter factor V to V_t and increase its coagulant activity (W. Liniger et al., [31]. The model is described by the reactions:

$$X_a + P \underset{k_{-1}}{\overset{k_1}{\rightleftharpoons}} X_aP$$

$$X_a + V_t \underset{k_{-2}}{\overset{k_2}{\rightleftharpoons}} X_aV_t$$

$$X_aP + V_t \underset{k_{-3}}{\overset{k_3}{\rightleftharpoons}} X_aPV_t$$

$$X_aV_t + P \underset{k_{-4}}{\overset{k_4}{\rightleftharpoons}} X_aPV_t$$

$$X_aPV_t \underset{k_{-5}}{\overset{k_5}{\rightleftharpoons}} X_a + T + V_t$$

$$V + T \underset{k_{-6}}{\overset{k_6}{\rightleftharpoons}} VT \quad \overset{k_7}{\underset{k_{-7}}{}} \quad T + V_t$$

With the use of the law of mass action this system is described by nine nonlinear differential equations giving the time-rate of change of concentrations of X_a, V, V_t, P, T, X_aP, X_aV_t, VT and X_aPV_t. These differential equations yield conservation laws for the enzyme X_a and its complexes, for prothrombin, thrombin and their complexes, and for the factors V, V_t and their complexes. Assuming the appropriate initial conditions and the usual steady state approximation and making a further simplification reduce rather complicated algebraic equations for the time rate r of formation of thrombin as function of the concentration [V] of factor V to:

$$r = \frac{A_1 + B_1[V] + C_1[V]^2}{1 + D_1[V]} \tag{7.1}$$

and as function of the concentration $[X_a]$ of factor X_a to:

$$r = \frac{A_2 + B_2[X_a] + C_2[X_a]}{1 + D_2[X_a]} \tag{7.2}$$

in which A_i, B_i, D_i (i = 1, 2) depend on the rate constants k_1 through k_{-7} and on the concentrations [P] and $[X_a]$, respectively [V] and [P].

Least square fitting of the observed data for r as function of [V] by Equation (7.1), while those of r as function of $[X_a]$ by Equation (7.2) leads to the Figures 7.5 and 7.6 respectively, in which the drawn curves are given by Equations (7.1) and (7.2) respectively. It is noteworthy that this mathematical model can accurately account for both maxima in the observed data shown in Figures 7.5 and 7.6 while the descriptive model could not account for those maxima. This, in addition to the good agreement between the observed data and computed results, shows clearly the great power of mathematical modeling of such complex phenomena.

The second mathematical model was proposed (W. Liniger, in preparation) to account for the activation of prekallikrein P substrate by the enzyme Hageman factor fragments, H, potentiated by the protein cofactor high molecular weight kininogen C. This model consists of the reactions:

$$H + P \underset{k_{-1}}{\overset{k_1}{\rightleftharpoons}} HP \xrightarrow{k_2} H + K$$

$$H + K \underset{k_{-2}}{\overset{k_2}{\rightleftharpoons}} HC$$

$$HP + C \underset{k_{-3}}{\overset{k_3}{\rightleftharpoons}} HPC$$

$$HC + P \underset{k_{-4}}{\overset{k_4}{\rightleftharpoons}} HPC$$

$$HPC \xrightarrow{k_5} H + K + C$$

Using the law of mass-action, these reactions are described by seven nonlinear differential equations which yield conservation reactions for the concentration h of the Hageman factor fragments, for the concentration p of the prothrombin, and for the concentration c of high molecular weight kininogen. Similarly, as in the first above-considered mathematical model it is found that the time rate r of formation of kallikrein is given by the equation

$$r = \frac{A + Bc + Cc^2}{1 + Dc + Ec^2} \tag{7.3}$$

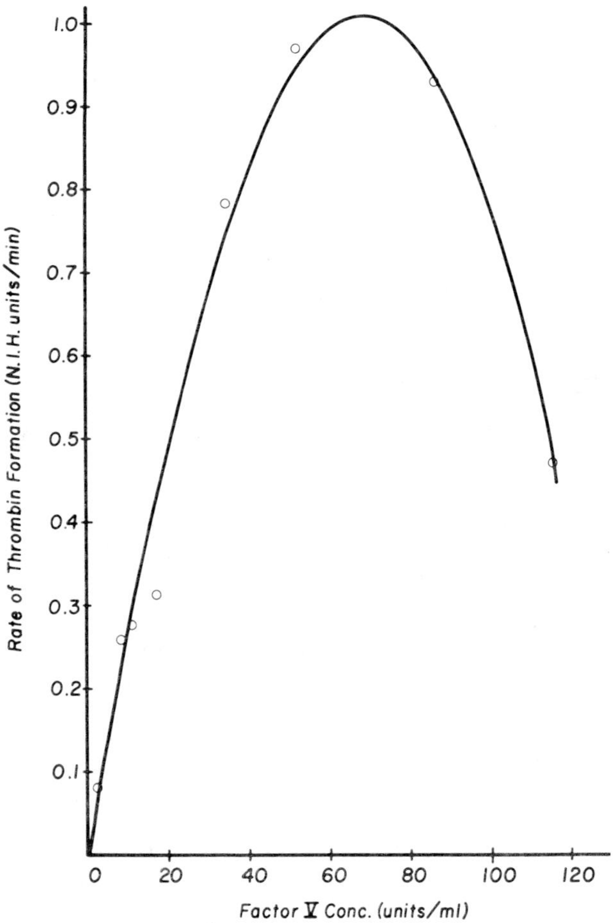

Fig. 7.5. The rate of thrombin formation as function of the concentration of factor V. The experimental data are given as open points and the curve is computed from Equation (7.1).

in which the coefficients A through E are functions of the rate constants and of p. The fitting of the experimental data with Equation (7.3), in a least square sense, is shown in Figure 7.7. Again, it is seen that the mathematical model locates the maximum of the curve accurately, for which the descriptive model could not account while the computed curve is also otherwise in agreement with the observational data.

In the absence of high weight molecular kininogen the system of reactions of this second model reduces to:

$$H + P \underset{k_2}{\overset{k_1}{\rightleftarrows}} HP \overset{k_2}{\rightarrow} H + K \qquad (7.4)$$

which is simply a Michaelis-Menten reaction scheme. Experimentally, as shown in Figure 7.8, the time-rate of formation on the prekallikrein concentration p is sigmoidal. W. Liniger (in preparation) has shown, by a singular perturbation method, that even such a simple reaction system as that of Michaelis-Menten can give such a sigmoidal dependence.

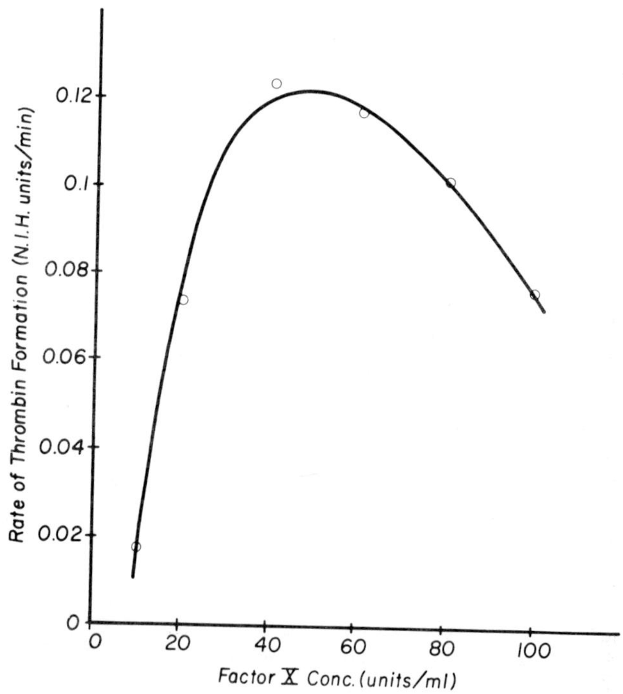

Fig. 7.6. The rate of thrombin formation as function of the concentration of factor X. The experimental data are given as open points and the curve is computed from Equation (7.2).

Recent experimental evidence has shown that in the last control subsystem surface effects play an important role. This, of course, is not unique for blood coagulation but also occurs in nucleation phenomena in physics (L. D. Landau and E. M. Lifshitz, [24]) as well as in biology (N. Rashevsky, [40] and in medicine, such as occurs in the formation of bubbles yielding "bends" which might lead to decompression sickness (G. Karreman, [20]). Nevertheless, the contribution of W. Liniger (in preparation) is significant as it shows that even such a very simple mechanism as leads to Michaelis-Menten kinetics may yield sigmoidal behavior.

As all biological organisms form open systems, not only reaction kinetics but diffusion is bound to occur in these complex systems. It is well known from relatively simple systems in which transformation of compounds occurs due to reactions, especially if they are autocatalytic, together with diffusion of those compounds, that bifurcations yielding changes in stability of states and limit cycle oscillations may occur in those systems (G. Nicolis and I. Prigogine, [35]). Hence, it is to be

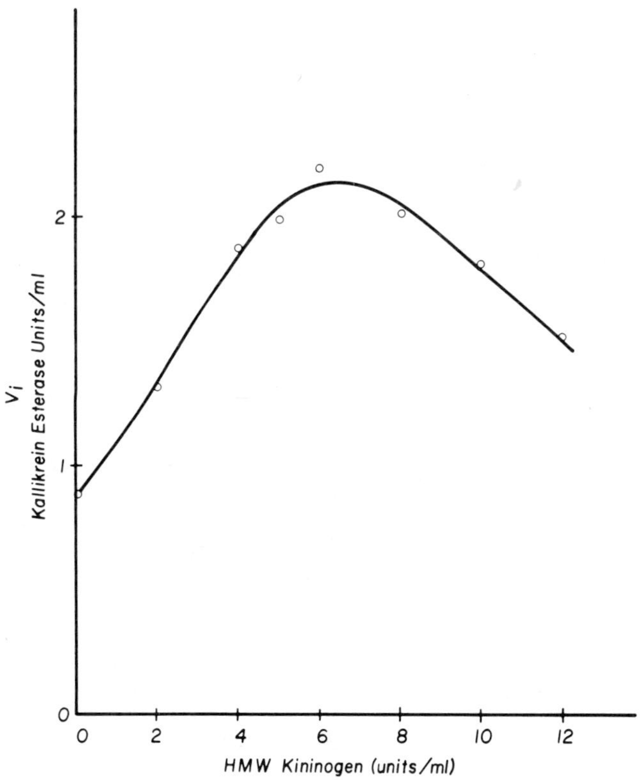

Fig. 7.7. The rate of kallikrein formation as function of high molecular weight kininogen. The experimental data are given as open points and the curve is computed from Equation (7.3).

expected, in fact it can be predicted, that several phase transitions and related all-or-none phenomena will be found in blood coagulation which consists of a very complicated hierarchy of cooperative cascade mechanisms. As already pointed out in the Introduction, fluctuations are very important for those phase transitions. One of the most fruitful ways to treat these problems mathematically is by a stochastic approach leading to a masterequation (H. Haken, [8]). The multidimensional character of these complicated systems can often be significantly reduced by the use of synergetics (H. Haken, [8], A. Pacault and C. Vidal, [36]). In this recently developed field, which has already found wide application to several sciences, use is made of the fact that in many practical applications only one or a very few modes become unstable. If all the other modes remain damped as is also often the case in practical situations, an adiabatec elimination procedure can be safely applied.

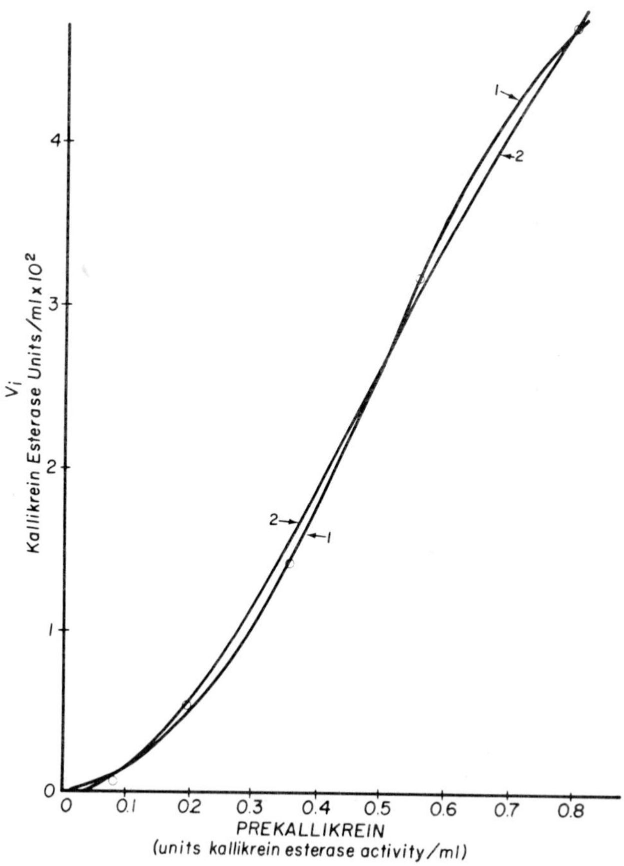

Fig. 7.8. The rate of kallikrein formation as function of the prekallikrein concentration. The experimental data are given by open points and the curve is computed on the basis of the reaction scheme (7.4).

Because all the damped modes follow the unstable modes, which take on the role of order parameters, the behavior of the whole system is often determined by the behavior of very few order parameters. As a consequence, even very complex systems may show a well regulated or (self-) organized behavior, determined by only a very few order parameters. The fundamental essence of this method is that the problem of instability is reduced to one of much smaller dimension than the original problem. The method is useful in that it allows the study of the impact of parameter changes on bifurcations. After the adiabatic elimination of the stable modes, equations are obtained for the order parameters which are then stabilized. If a system is completely deterministic, it cannot jump by itself from one stable state to another, still more stable, state. It will require an initial force or randomly repeated forces to transfer the system from a stable state to a still more stable state. In all practical cases of natural systems such random forces do occur due to fluctuations, in laser physics due to spontaneous emission of light, in biological evolution due to mutations and in threshold phenomena due to random noise. The recent great progress in the study of critical phenomena is bound to have several spin-offs in many other fields of science. The importance of scale invariance is very fruitful for the explanation of critical phenomena (F. J. Wegner, [53], S. K. Ma, [32], M. S. Green, [5]). This scale invariance is based on the idea of a fixed point hamiltonian which is invariant under a change of the length scale (F. J. Wegner, [53]). The introduction of the renormalization group approach to scaling by K. G. Wilson, ([54], [55]) has already shown great usefulness. It has been shown that the exponents of certain power laws have a universal character in critical phenomena (S. K. Ma, [33]). Invariance to time reversal has also been shown to be important for Schrödinger dynamics. The above-outlined important progress in the theory of cooperative processes, which has already given much insight for the description and explanation of critical phenomena, will probably produce significant progress for the explanation of cooperative phenomena in biology, like physiological excitation, and in medicine, like blood coagulation.

As already emphasized above, the processes occurring in blood coagulation are cooperative in nature. The essential feature of cooperative phenomena is (nearest) neighbor interaction, as is well known from physics, in which ferromagnetism and alloy-formation have long been the prototype. G. N. Ling (loc. cit, [26] and in preparation) has been able to explain many biological phenomena, such as occur in the uptake of cations like potassium and sodium, on the basis of his association-induction hypothesis. On the basis of such cooperative mechanisms (G. Karreman, [16]) it has been possible to derive a cooperative specific absorption isotherm which has been found to be in agreement with many experimental data (A. W. Jones and G. Karreman, [10]; G. N. Ling and G. Bohr, [27]; G. Karreman, [18]). In addition, the solution of the masterequation for such cooperative specific absorption phenomena (G. Karreman, [17]) has been found to be in good agreement with the rate of potassium-sodium exchange by human lymphocytes (W. Negendank and G. Karreman, [34]). It has already been found possible to obtain the right order of magnitude of the threshold of physiological excitation (G. Karreman, [19]). Further studies along these lines, taking into account the changes in the assembly of water molecules cooperatively absorbed at protein backbone sites (G. N. Ling, [28]), due to chemical induction as a consequence of a stimulus, with resulting changes in bioelectric potentials are in progress (G. Karreman, in preparation). These bioelectric studies are also important for blood coagulation as factor XII is bound to a negative charged surface.

The above-mentioned chemical induction process is basically a wave mechanical phenomenon. The application of wave mechanics to biological problems is still, in spite of some recent important progress (H. F. Schaefer, [46], [47]), largely in its initial stages.

The general significance of cascade mechanisms is much more widespread in science than their occurrence in blood coagulation and other biological processes. Energy production by nuclear reactions involves cascades in which nuclear particles play an important role. The cascades involved in these reactions are essentially self-sustaining neutron reactions. Possibly other nuclear cascade mechanisms may merit consideration. It is known that the probability for effective nuclear transmutation of N^{7+} nuclei with alpha particles producing nuclei of the oxygen isotope 17 and a proton is a small quantity. However, by combining this fusion reaction with such fusion reactions in which protons bombard lithium isotope 17 producing <u>two</u> alpha particles, hence leading to a cascade system with amplification may yield, with the use of nuclear accelerators such as cyclotrons and synchrotons, potentially new nuclear energy sources if appropriate safety measures are taken against radioactivity. Other possible new energy sources can potentially be developed by cascade mechanisms using biochemical reactions (or their simulations), especially if they are initiated by solar energy like photosynthesis, involving enzymes or catalysts with high turnover rates and fast transfer reactions as, for instance, due to the involvement of free radicals produced by charge transfer complexes as in the recent theory of cancer proposed by A. Szent-Györgyi ([50]).

REFERENCES:

1. Andronov, A. A. et al., Theory of Bifurcations of Dynamic Systems on a Plane, Halsted, New York, 1973.

2. Del Re, G. and Ladik, J., Absorption of molecules on polymers, I. A general analysis in terms of the green matrix version of the local impurity SCF approach, J. Chem. Phys. In press.

3. Ginsburg, V. L. and Landau, L. D. in Quantum Theory of Many Particle Systems, (by A. L. Fetter and J. D. Walecka), McGraw-Hill, New York, 1971.

4. Glansdorf, R. and Prigogine, I., Thermodynamic Theory of Structure, Stability and Fluctuations, Wiley Interscience, New York, 1971.

5. Green, M. S., Invariance Properties of the Renormalization Group, in Critical Phenomena (J. Brey and R. B. Jones, eds.), Springer, New York, 1976.

6. Gurel, O., Dynamics of cancerous cells, Cancer 23: 497 (1969).

7. Gurel, O., Bifurcation models of mitosis, Physiol. Chem. and Physics 4: 139 (1972).

8. Haken, H., Synergetics: An Introduction Nonequilibrium Phase Transitions and Self-Organization, in Physics, Chemistry and Biology, Springer, New York, 1st Ed. 1977, 2nd Enlarged Ed., 1978.

9. Haken, H., (ed.), Pattern Formation by Dynamic Systems and Pattern Recognition, Springer, New York, 1979.

10. Jones, A. W. and Karreman, G., Ion exchange properties of the canine carotid artery, Biophys. J. 9: 884 (1969).

11. Karreman, G., et al., On the mechanism of action of chlorpromazine, Science 130: 1191 (1959).

12. Karreman, G., Contributions to quantum biology, I. Mobile electronic characteristics of riboflavin radicals, Bull. Math. Biophys. 23: 55 (1961a).

13. Karreman, G., Studies in quantum biology, II. The mobile electron characteristics of tryptophan$^+$ in relation to those of FMN^-, FMNH and $FMNH_2^+$, Bull. Math. Biophys. 23: 135 (1961b).

14. Karreman, G., Electronic aspects of quantum biology, Ann. N. Y. Acad. Sci. 96: 1029 (1962).

15. Karreman, G., Electronic Features of Some Psychopharmacological Compounds, in Data Acquisition and Processing in Biology and Medicine, Pergamon Press, New York, 1962.

16. Karreman, G., Cooperative specific absorption of ions at charged sites in an electric field, Bull. Math. Biophys. 27: 91 (1965).

17. Karreman, G., Stochastic treatment of cooperative specific absorption, Bull. Math. Biophys. 33: 483 (1971).

18. Karreman, G., Cooperative specific absorption, Ann. N. Y. Acad. Sci. 204: 393 (1973).

19. Karreman, G., Towards a physical explanation of physiological excitation as a cooperative specific absorption phenomenon, Bull. Math. Biol. 35: 149 (1973).

20. Karreman, G., Physical Aspects of Bubble Formation in Tissues, Proc. Fifth Symposium on Underwater Physiology (C. J. Lambertsen, ed.), p. 293, FASEB, Bethesda, Maryland, 1976.

21. Ladik, J. J., Towards the Theoretical Determination of the Conformation of Biological Macromolecules, in Quantum Science (J. L. Calais et al., eds.), 1976.

22. Ladik, J. J., Electronic Structure of Biopolymers, in Quantum Theory of Polymers (J. M. Andre et al., eds.), D. Reidel Pub. Co., Dordrecht, 1978.

23. Ladik, J. J., Electronic Structure of Proteins and DNA: Solid-State Aspects, Submolec. Biol. and Cancer, Ciba Foundation Series 67 (new series), Excerpta Medica, 1979.

24. Landau. L. D. and Lifshitz, E. M., Statistical Physics, Pergamon Press, Oxford, 1958.

25. Ling, G. N., A Physical Theory of the Living State: The Association-Induction Hypothesis, Blaisdel Publishing Company, New York, 1962.

26. Ling, G. N., A new model for the living cell; a summary of the theory and recent experimental evidence in its support, Int. Rev. Cytol. 26: 1 (1969).

27. Ling, G. N. and Bohr, G., Studies on ion distribution in living cells, II. Cooperative interaction between intracellular potassium and sodium ions, Biophys. J. 10: 519 (1970).

28. Ling, G. N., Hydration of Macromolecules, in Water and Aqueous Solutions (R. A. Horne, ed.), Wiley, New York, 1972.

29. Ling, G. N. et al., Mg^{++} and K^{+} distribution in frog muscle and egg: a disproof of the Donnan theory of membrane equilibrium applied to the living cells, J. Cell. Physiol. 101: 261 (1979).

30. Ling, G. N., The Polarized Multilayer Theory of Cell Water and Other Facets of the Association-Induction Hypothesis Concerning the Distribution of Ions and Other Solutes in Living Cells, in The Aqueous Cytoplasm (A. D. Keith, ed.), Marcel Dekker, New York, 1979.

31. Liniger, W. et al., Mathematical model of the activation of prothrombin by factor X_a and factor V_t, Bull. Math. Biol. In Press.

32. Ma, S. K., Scale Transformations in Dynamic Models, in Critical Phenomena (J. Brey and R. B. Jones, eds.), Springer, New York, 1976.

33. Ma, S. K., Modern Theory of Critical Phenomena, Benjamin, Reading, Mass., 1976.

34. Negendank, W. and Karreman, G., Rate of potassium-sodium exchange by human lymphocytes: prediction of the cooperative absorption model, J. Cell. Physiol. 98: 107 (1978).

35. Nicolis, G. and Prigogine, I., Self-Organization in Nonequilibrium Systems from Dissipative Structures to Order Through Fluctuations, Wiley Interscience, New York, 1977.

36. Pacault, A. and Vidal, C., eds., Synergetics Far from Equilibrium, Springer, New York, 1979.

37. Poincaré, H., Les Méthodes Nouvelles de la Mécanique Céleste, Vol. 1, Gauthier-Villars, Paris, 1892.

38. Pullman, A. and Pullman, B., Cancérisation par les Substances Chimiques et Structure Moléculaire, Masson & Cie., Paris, 1955.

39. Pullman, B. and Pullman, A. Quantum Biochemistry, Wiley Interscience, New York, 1963.

40. Rashevsky, N., Physico-Mathematical Foundations of Biology, Vol. 1, Dover Publ., New York, 1960.

41. Rashevsky, N., Physico-Mathematical Foundations of Biology, Vol. 2, Dover Publ., New York, 1960.

42. Rosen, R., Morphogenesis, Chapt. 1 in Foundations of Mathematical Biology (R. Rosen, ed.), Acad. Press, New York, 1972.

43. Rosen, R., Mechanics of Epigenetic Control, Chapt. 2 in Foundations of Mathematical Biology (R. Rosen, ed.), Acad. Press, New York, 1972.

44. Rosen, R., Some Relational Cell Models: The Metabolism-Repair Systems, Chapt. 4 in Foundations of Mathematical Biology (R. Rosen, ed.), Acad. Press, New York, 1972.

45. Rosen, R., Fundamentals of Measurement and Representation of Natural Systems, Elsevier North Holland, New York, 1978.

46. Schaefer, H. F., Methods of Electronic Structure Theory, Plenum Press, New York, 1977.

47. Schaefer, H. F., Applications of Electronic Structure Theory, Plenum Press, New York, 1977.

48. Szent-Györgyi, A., Towards a new biochemistry, Science, June 27, 609: (1941).

49. Szent-Györgyi, A., The study of energy levels in biochemistry, Nature 148: 157 (1941).

50. Szent-Györgyi, A., Electronic Biology and Cancer: A New Theory of Cancer, Marcel Dekker, New York, 1976.

51. Thom, R., Stabilité Structurelle et Morphogénèse Essai d'une Théorie Générale des Modèles, Benjamin, Inc., Reading, Mass., 1972.

52. Waddington, C. H., Towards a Theoretical Biology, Vol. 1, Univ. Press, Edinburgh, 1968.

53. Wegner, F. J., Critical Phenomena and Scale in Variance, in Critical Phenomena (J. Brey and R. B. Jones, eds.), Springer, New York, 1976.

54. Wilson, K. G., Renormalization group and critical phenomena I, Phys. Rev. B4: 3174 (1969).

55. Wilson, K. G., Renormalization group and critical phenomena II, Phys. Rev. B4: 3184 (1969).

56. Woodcock, A. E. R., and Poston, T., A Geometrical Study of the Elementary Catastrophes, Springer, New York, 1974.

57. Zeeman, E. C., Catastrophe Theory, Addison-Wesley Publ. Co., Reading, Mass., 1977.

CURRENT MODELS OF POSITIONAL INFORMATION

Stuart A. Kauffman
Department of Biochemistry and Biophysics, University of Pennsylvania, School of Medicine, Philadelphia, Pennsylvania

INTRODUCTION: POSITIONAL INFORMATION

The past decade has witnessed a renewal of deep interest in the problem of pattern formation in developmental biology. In large measure, the resurgence of enthusiasm coincides with Wolpert's reformulation of this fundamental problem in terms of the concept of positional information ([87], [88]). Novel features of Wolpert's theory have led both to proposals of alternative "coordinate systems" supplying positional information, and to a rich variety of experiments designed to test the alternatives. The present article provides a brief review of the major alternatives which leads to the formulation of a new class of models based on the unrecognized, but common capacity of biochemical reaction-diffusion systems to generate transverse (cross) gradients in growing asymmetrical tissue domains.

Among the important experimental techniques used to study pattern formation are those which rely on surgical intervention, often coupled with grafting, and analysis of subsequent pattern regulation. Two broad classes of pattern regeneration have long been distinguished; _morpholactic_, in which a truncated tissue regenerates within its bounds the full range of pattern elements normally located in the entire tissue; and _epimorphic_, in which a truncated tissue forms a regeneration blastema which undergoes cell proliferation enlarging the initial truncated fragment, and reforms missing pattern elements among the new cells of the regeneration blastema (Morgan [55], reviewed by Slack [72]). This article focuses on epimorphic regulation.

Prior to Wolpert's introduction of the positional information concept, the dominant theory available postulated the existence of developmental fields [85] possessing "prepatterns" ([75], [82]), non-uniform spatial distributions of biochemical substances in a tissue, whose local peaks would induce the formation of pattern elements such as specific digits, sensillae, or bristles.

In contrast to Stern [75], Wolpert ([87], [88]) proposed the more abstract idea that cells within a developmental field possess positional information about their locations with respect to the boundaries of the field, through access to an underlying positional coordinate system. The behavior of each cell in the field was assumed to be due to two independent processes. The cell _assesses_ its position, then interprets its positional information according to the type of cell it is, and forms a specific structural element in the overall pattern.

The chief difference between Wolpert's positional information and Stern's prepattern, is that positional information is free of assumptions about the existence of specific morphogen peaks underlying the subsequent differentiation of specific pattern elements. This freedom in one sense makes the theory of positional information less predictive, yet allows for two important possibilities; first, that the positional information system in all the developmental fields of one organism are identical; and more radically, that the positional information system in all organisms, or all epimorphic fields, is identical.

Strong evidence supports the more restricted hypothesis that a single system of positional information may underlie all the developmental fields of a single organism. A number of well studied homeotic mutants [59] have been found in the fruit fly Drosophila melanogaster. Homeotic mutants replace part or all of one adult ectodermal structure with another. For example, Aristapedia [64] and Nasobemia [32] convert part or all of the antenna to a mesothoracic leg, tumorous head converts part or all of the head to posterior abdomen and genitalia, eyeless ophthalmoptera [61] converts eye to wing. A particularly striking observation in such homeotic tissue is that even if a very small patch of distal antenna is converted to leg, it is converted to distal leg. Conversely, if proximal antenna is converted to leg, it forms proximal leg ([62], [59]). Analysis of such partially transformed appendages leads to serious difficulties for a prepattern model. If even a small adult patch derived from less than a score of cells forms proximal leg bristles in the proximal antenna, then either the small group of preleg cells forms its own small prepattern with appropriate morphogen peaks or the prepattern in the antenna is not specific to the antenna, but a more general appendage prepattern whose peaks no longer correspond to specific positions of antenna, leg, genital or wing pattern elements. The former interpretation has seemed to most workers improbable, while the latter shades smoothly into the concept that the prepattern is some system supplying positional information subsequently interpreted by cells. Hence, if the positional information in an antenna and leg are identical, proximal antennal cells in the antennal tissue will interpret that information and form appropriate antenna pattern elements, while neighborhing homeotically transformed leg cells will interpret the same information and form appropriate proximal leg structures [87].

The general success of the positional information concept led to a search for the coordinate system which supplies positional information. At present, polar ([10], [14], [31]), spherical [66], Cartesian ([47], [48], [72], [86]), and general orthogonal [24] models have been proposed. Differences among these alternatives are far from trivial. Although it is always possible mathematically to transform from one to another coordinate system, the "forces" or tissue properties which must be postulated to explain the observed features of pattern regulation differ sharply in the different models. While the morphogens have yet to be found, one task in this area of biology consists in efforts to discover the coordinate system and "forces" or requisite cell properties which most simply account for the observations, with the hope that the proper formulation will provide both macroscopic laws at the tissue level, and aid in discovery of the underlying molecular variables.

THE PHENOMENA

To assess the relative success of the alternatives which have been proposed, it is necessary to review briefly at least some of the major phenomena of pattern formation and regeneration.

1) Intercalary Regeneration of Intervening Structures

If an amphibian limb capable of regeneration is transected proximal and distal to the elbow, and the distal wrist fragment grafted to the proximal shoulder stump, cell proliferation forms in the wound area, followed eventually by regeneration of the missing elbow region. This basic phenomenon is fairly ubiquitous ([56], [71], [72], [76]) and fundamental. Juxtaposition of normally non-adjacent tissues from a single development field is generally followed by regeneration, in proper spatial order, of the structures normally lying between the juxtaposed tissue edges. This notion of "betweenness" is necessarily central to any theory of pattern formation.

The simplest physical model to account for "betweenness" in intercalary regeneration posits the existence of one or more chemical concentration gradients spanning the tissue, whose concentration levels specify the positional information of cells at each point in the domain. As shown in Figure 8.1, in which a proximal distal gradient along an amphibian limb is envisioned, surgical removal of the elbow and grafting of wrist to shoulder creates a discontinuity in the gradient, [S], at the

Fig. 8.1. Gradient in concentration of [S] provides proximal-distal positional information in amphibian limbs. Serial threshold levels specify pattern elements A, B, ... I. Removal of mid region of limb, D, E, F, and grafting, creates discontinuity, which stimulates cell proliferation. Diffusion smoothing of gradient discontinuity regenerates missing gradient levels (wavy lines), and structures DEF.

graft junction. If one imagines that gradient concentrations remain fixed in the "old" tissue fragments, while diffusion occurs in the new cells of the wound blastema, then simple diffusive averaging of the concentration discontinuity at the graft junction smooths over the discontinuity, recreating all the intervening gradient values in proper spatial order.

For the remainder of this review, I shall adopt the postulate that position is specified by graded scalar properties in tissues, although it is important to stress that alternative discrete models [1], have been formulated. Given the postulate of positional gradients, a fundamental question is the extent to which the simple property of diffusive-like averaging of gradient discontinuities can account for the phenomena of pattern formation and regeneration. This simple property turns out to be sufficiently powerful to account for the bulk, but not all, of the available data.

2) Sequential Formation of Positional Axes in Development

In several systems, positional axes appear to be established sequentially during development. In classical experiments, Harrison ([36], [37]) removed the right forelimb bud of Amblystoma and grafted in its stead the left forelimb bud. Such grafts must invert either the anterior-posterior limb axis while keeping donor and host dorsal and ventral axes aligned; or invert dorsal-ventral donor and host axes, while keeping anterior and posterior axes aligned. Harrison found that if very early left limb buds were grafted to the right, they developed into normal right limbs. If late left limb buds were grafted, they formed normal left limbs with that axis inverted with respect to the host that was inverted at surgery. But if left to right grafts were made at an intermediate stage, the outcome depended upon which axis was inverted at the graft junction. If the anterior-posterior axis remained normally aligned and the dorsal-ventral axis was inverted, the donor left limb bud formed a right limb, while if the dorsal-ventral axis remained aligned and the anterior-posterior axis was inverted at the graft junction, the donor left limb bud formed a left limb which remained inverted at the host donor junction. Harrison interpreted his results to imply that the donor anterior-posterior axis becomes automatically self sustaining prior to the dorsal-ventral axis. Similar data suggest that amphibian eye ([44], [45]) and ear ([38]) axes, as well as limb axes are established sequentially, although the status of data on the eye is in dispute.

In a variety of insects, the available evidence suggests that the longitudinal body plan of the entire embryo becomes established prior to fixation of dorsal ventral commitments (reviewed by Sander [67]).

3) Distal Transformations

If an amphibian limb or tail is transected at the elbow, the proximal stump can form a regeneration blastema and regenerates the distal limb [36]. If the digits of the transected distal fragment are implanted into a host flank to establish an adequate blood supply to the distal fragment, the cut surface at the elbow, which initially faced proximally, forms a regeneration blastema and regenerates a second distal wrist and hand structures which are mirror symmetrical to the initial implanted distal limb ([19], [20], [26], [36]). In these experiments, both the proximal stump and the implanted distal limb fragment regenerate the same set of distal limb structures from the cut surfaces at the elbow, identified as a regenerate in the proximal stump and duplicate on the implanted distal limb fragment. The fact that both fragments form distal limb has been called the rule of distal transformation [65]. Similar results have been found in many insect legs ([5], [16], [17], [30]) and in the imaginal discs of *Drosophila* ([31], [40]).

4) Supernumerary Limbs

Among the most striking observations in pattern regulation is the induction of extra or supernumerary limbs following grafting operations. After both the anterior-posterior and dorsal-ventral amphibian limb axes are fixed, transplantation of a left distal limb to a right proximal stump which reverses the anterior-posterior axis of donor relative to host but leaves the dorsal-ventral axis aligned, typically results in formation of two supernumerary limbs at the anterior and posterior margins of the donor host junction. If instead the anterior-posterior axes of host and graft are aligned, but dorsal-ventral axes are inverted, the two supernumeraries emerge from the dorsal and ventral margins of the host donor junction. These supernumerary limbs generally have the handedness of the proximal stump ([36], [37], [15]). Similar results have been obtained in transplantation of cockroach limbs ([5], [8], [16], [30]).

Rotation of a left distal limb by 180° and regrafting to its own proximal stump results in a more variable range of results. After such rotations, the donor may partially rotate back toward its normal alignment; sometimes no, one, two, or more supernumerary limbs are formed at the graft site, with the same or opposite handedness ([5], [8], [15], [16], [31], [54], [80]).

6) Duplication and Regeneration by Complementary Fragments of a Tissue

Distal transformation by both proximal and distal limb fragments of amphibians is one example of duplication and regeneration by complementary fragments of a developmental field. The phenomenon is common, however, and has been studied in greatest detail in the imaginal discs of Drosophila melanogaster. The bulk of the remaining discussion of data pertaining to pattern regulation will focus on this system.

DROSOPHILA IMAGINAL DISCS

Drosophila melanogaster is a holometabolous insect with egg, larva, pupa and adult stages. During metamorphosis, the larval ectoderm lyses, and the ectoderm of the adult is formed by the terminal differentiation of special larval organs called imaginal discs [33]. In the late third instar larva, each imaginal disc is a two dimensional sheet of cells forming the surface of a hollow sphere. The columnar cells on one hemisphere form the imaginal disc proper, while the thin squamous cells on the remaining hemisphere form the peripodal membrane which is lost during metamorphosis [33]. Imaginal discs are found as bilaterally symmetric pairs, each destined to form specific left and right regions of the adult ectoderm: the left and right first leg discs form the two prothoracic legs; the two wing-thorax discs form the left and right mesothoraces and wings, etc.

By injecting specific fragments of each disc into host larvae for metamorphosis it has been possible to construct a fate map of each type of imaginal disc ([9], [11], [69], [84]). For example, the fate map of the wing-thorax (hereafter wing) disc, Figure 8.2, shows that the upper and lower margins of the disc along its longitudinal axis form ventral and dorsal thoracic structures, while the mid region of the disc forms wing structures. During metamorphosis, the wing disc folds along a bent arc running from the anterior to posterior disc margin, and apposes ventral and dorsal thorax areas, ventral and dorsal wing hinge areas, and ventral and dorsal wing blade areas, creating a "bag" which everts through the peripodal membrane. The center of the disc maps to the distal wing tip, while an arcing line from anterior to posterior disc edge through the disc center corresponds to the wing margin.

Grafting experiments like those in amphibian or cockroach limbs are not yet feasible in Drosophila. However, closely analogous experiments can be performed by cutting the wing disc into known fragments, and injecting each fragment into the abdomen of a fertilized adult female. In that environment, the disc fragment heals its cut edge, apposing tissue regions which are normally non-adjacent [63] and new cells form in the wound area. After a week in culture, the mass of a disc fragment typically doubles. After culture in adult abdomens, fragments may be recovered and injected into host larvae for metamorphosis, then recovered from the emerged host adult. By comparison of the patterns of hairs, sensillae and bristles formed when a known disc subfragment is injected directly into larvae for immediate metamorphosis, it is possible to characterize the pattern regulation which occurs in the cultured fragment.

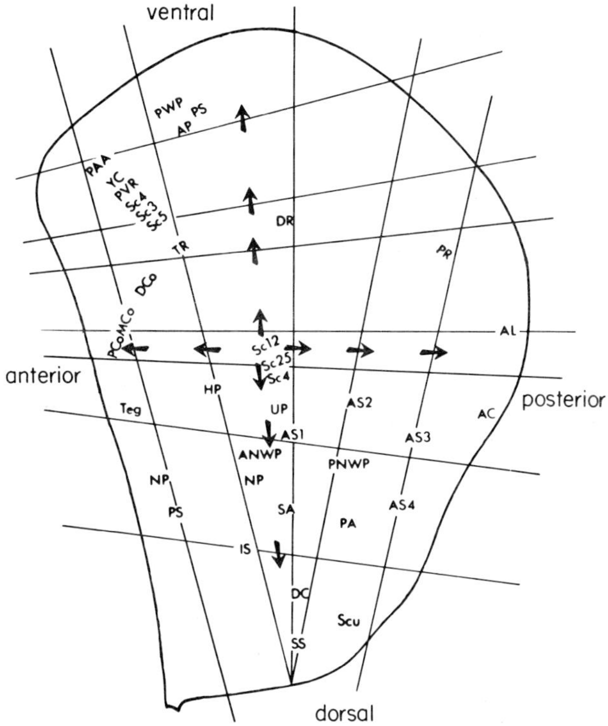

Fig. 8.2. Fate map of Drosophila wing disc. Lines show positions of single cuts. Arrow across each line points from the fragment created by that cut which regenerates, into the fragment which duplicates. From Bryant [9]. Abbreviations for wing disc pattern elements: PST, presutural bristle; NP, notopleural bristles; SA, supraalar bristles; PA, postalar bristles; DC, dorsocentral bristles; Scu, scutellar bristles; ANWP, PNWP, anterior and posterior notal wing processes; Teg, tegula; HP, humeral plate; UP, unnamed plate; AS1-4, first to fourth axillary sclerites; Pco, Mco, Dco, proximal, medial and distal costa; TR, triple bristle row (anterior wing margin); DR, double bristle row (distal wing margin); PR, posterior row of hairs; Sc4, Sc25, Sc12, groups of sensilla campaniformia on the proximal dorsal radius; AL, alar lobe; AC, axillary cord; PAA, prealar apophysis; YC, yellow club; PVR, proximal ventral radius; PWP, pleural wing process; PS, pleural sclerite; AP, axillary pouch; Sc4, Sc3, Sc5, groups of sensilla campaniformia on the proximal ventral radius; Anterior, Posterior Ventral and Dorsal disc margins.

The following are the dominant results:

a. If the wing disc is cut in two fragments by a straight cut, the roughly smaller fragment duplicates some or all of its structures, in favorable cases yielding a mirror symmetrical duplicate whose symmetry line lies near the cut on the wing disc; the larger complementary fragment regenerates structures normally formed by its smaller complement [9]. Thus complementary fragments often exhibit complementary behavior, one duplicating, one regenerating. Consequently, it is possible to draw an arrow across each single straight cut on the disc, pointing from the fragment which regenerates into that which duplicates. As shown in Figure 8.2 such arrows point radially outward from a small region in the interior of the disc. The polarity of regeneration and duplication reverses about this interior point, termed the "high point."

b. If the disc is cut into arbitrary 3/4 and 1/4 "pie" sectors, the 3/4 fragment regenerates, the 1/4 fragment duplicates [10].

c. If an interior "distal" circular region containing the "high point" is cut and cultured, it duplicates, the duplicate forming a mirror symmetric hemisphere to the original high point region. If the outer "proximal" annulus remaining after the high point region is removed is cultured, the outer annulus regenerates the central "distal" high point region ([9], [40]).

d. If two narrow normally duplicating crescent fragments from opposite edges of the disc are mixed, they regenerate the intervening pattern elements spanning across the disc [39].

e. Complementary fragments do not strictly duplicate and regenerate. Under different experimental conditions, and with varying frequencies normally duplicating fragments can regenerate most of the pattern elements in the complementary fragment ([9], [27], [40], [48], [51]).

f. Anterior-posterior and ventral-dorsal asymmetries exist in the capacity of narrow duplicating fragments to regenerate: narrow anterior margin and ventral thorax margin structures have been found by a number of workers to regenerate extensively, while narrow posterior and dorsal margin fragments do not ([27], [40], [48]). The results in other imaginal discs are fundamentally similar ([11], [69], [84]), although an interior "high point" has not been demonstrated.

g. Mutants causing pattern duplication in different imaginal discs have been analyzed by a number of workers ([12], [46]) and will be discussed further below.

MODELS OF POSITIONAL COORDINATE SYSTEM

The first major advance in predictive use of the Positional Information hypothesis lay in the formulation of the polar coordinate, or "clockface" model for pattern regulation in epimorphic fields by French *et al*. [31]. The model was based on some of the results described in amphibian limbs, cockroach limbs, and imaginal discs, and is well illustrated by application to the wing of *Drosophila*.

The existence of a "high point" in the central (distal) region of the disc, Figure 8.2, about which the direction of regeneration reverses suggested that cells might measure their distance in the tissue from this special point. This raised the possibility that the position of cells is specified in a polar coordinate system with the high point as its origin. Since the wing disc is a two dimensional surface, azmuthal angle must be measured. Were angle specified by a single scalar variable, that variable would necessarily be discontinuous along some radial line in the tissue; for example along the radial line $\phi = 2\pi = 0$. But Bryant found that any 1/4 pie wedge fragment of the wing disc duplicates, while its 3/4 complement regenerates. If an azimuthal discontinuity were present, the 1/4 wedge fragment containing it should behave differently from the rest, and regenerate. Failure to find evidence of an angular discontinuity implies that, if cells measure angle, they do so seamlessly. Therefore the model proposes that cells measure radial distance from a distal "high point" origin, and angle seamlessly modulo 2π.

In order to account for the bulk of the data on epimorphic regeneration and duplication, the polar coordinate model initially proposed two rules of intercalary regeneration, and a third special rule for distal transformation.

Rule 1. If cells having different radial values are apposed, cell proliferation will be stimulated, and the missing intervening radial values will be restored back to a resting radial gradient, then proliferation will cease.

Rule 2a. If cells having different angular values are apposed, cell proliferation will be stimulated and the missing angular values will be intercalated back to a resting angular gradient.

Rule 2b. Since two angular arcs around a 2π circle of values join any two juxtaposed angular values, a choice rule is needed. The simplest postulates that angular intercalation occurs along the shorter arc.

Special Rule 3, the complete circle rule. If a complete circle of angular values at a proximal radial level is exposed, distal regeneration occurs. The special nature of Rule 3 will be discussed below.

The two intercalation rules 1 and 2a, b, suffice to explain major features of epimorphic pattern regulation. For example, the classical example of grafting an amphibian hand to shoulder with regeneration of the missing elbow region is explained by rule 1. Proximal radial values in the shoulder apposed to distal radial values in the hand lead to intercalation of the intervening radial values specifying elbow.

Duplication and regeneration by complementary fragments of the wing disc is explained by rules 2a, b. Figure 8.3 shows a single straight cut on the wing disc, yielding a narrow anterior fragment and a large posterior fragment. During culture, the narrow anterior fragment folds over, apposing the cut edge such that the ventral and dorsal thoracic regions heal together, ventral and dorsal wing blade regions

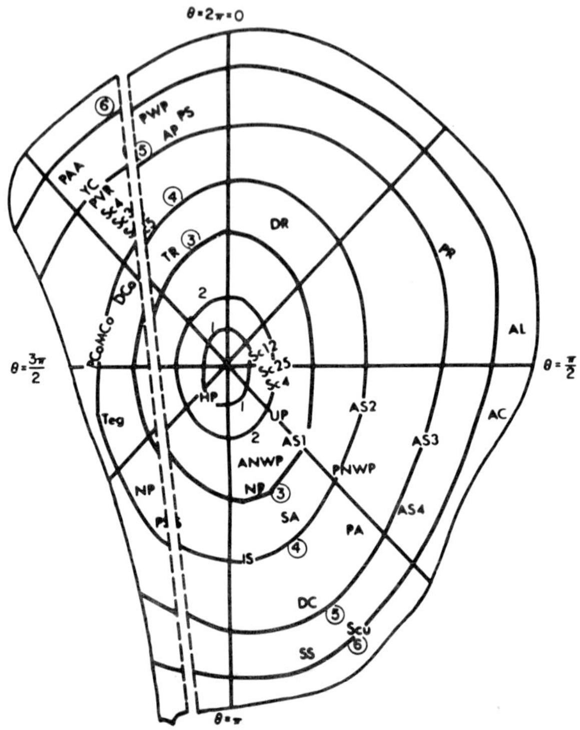

Fig. 8.3. The polar coordinate model of the wing disc. Radial values (1-6) are measured from the "high point." Angle, ϕ, is measured without discontinuity modulo 2π. A straight cut anterior to the high point creates a narrow anterior fragment and wide posterior fragment. The latter heals by folding along its wound margin, apposing positions with equal radial but discordant angular values. Cell proliferation and smoothing of angular discontinuities along the short arc between apposed values leads to regeneration by the posterior fragment. The narrow anterior fragment folds and heals along its wound margin, juxtaposing the same positional values as did the posterior fragment. Short arc intercalation leads to duplication by the anterior fragment.

also heal together [63]. This healing juxtaposes cells with similar radial values, but discordant angular values. This discontinuity stimulates cell proliferation, and smoothing of the angular discontinuity along the shorter angular arc. Since this shorter angular arc is the arc present in the original narrow anterior fragment, the positional values in the new cells form a mirror symmetrical duplicate of those in the original anterior fragment, and the fragment duplicates.

The positional values present along the cut margin of the large posterior fragment are identical to those along the cut margin of the narrow anterior fragment. Therefore, if the large fragment wound heals in a similar way, the same pairs of positional values must be apposed in the posterior fragment as in the anterior fragment.

Therefore, the posterior fragment must intercalate, along the shorter angular arcs, the same intervening positional values as did the anterior fragment. Therefore, the posterior fragment regenerates.

The polar model demonstrates a more general result. Whatever the coordinate system specifying position in a developmental field may be, the positional values along the two margins of a single cut are identical. If the two fragments heal in similar ways, both will appose essentially identical pairs of positional values. Therefore, if subsequent pattern regulation is governed by diffusive-like averaging of positional disparities, both complementary fragments must reform the <u>same</u> set of structures. If one fragment duplicates, the second must regenerate. The prediction of complementary behavior in complementary fragments is a coordinate free property which follows from the postulates of graded positional values, and averaging of disparities.

Since the polar model is symmetric about the "high point," a narrow posterior fragment will duplicate, its complement will regenerate. Therefore the direction of regeneration will reverse about the high point. Further, any 1/4 wedge section will heal its two cut margins and duplicate, while its 3/4 complement will regenerate.

INTERCALARY REGENERATION, BETWEENNESS, AND CONVEX SETS

A central feature of the postulates of intercalary smoothing is that only those positional values lying <u>between</u> the apposed values can be reformed. This leads to a critical restriction in the predictive consequences of any given coordinate system, since it implies that diffusive-like smoothing can <u>only</u> recreate positional values lying in the <u>convex set</u> bounded by the positional values in the apposed tissue edges. This restriction, in turn, implies that different coordinate systems may demand different special cellular behaviors beyond simple diffusive-like smoothing to account for the data.

The concept of a convex set, and the limitations it imposes can be brought out in the polar coordinate model. Radial position can be visualized without loss of generality as a radially symmetric gradient whose peak is at the distal "high point." In Figure 8.4, I show a wing disc from which the distal high point region has been removed, thus removing the radial gradient "peak." In the remaining outer proximal annulus, only lower values of the radial gradient are present. Therefore, no juxtaposition of tissue edges in the proximal annulus can lead to "diffusive-like" filling in of the missing high point radial peak. In a polar model, if the region containing the origin is removed, that region does not lie "between" the positional values in the proximal annulus. That is, the region around the origin is not in

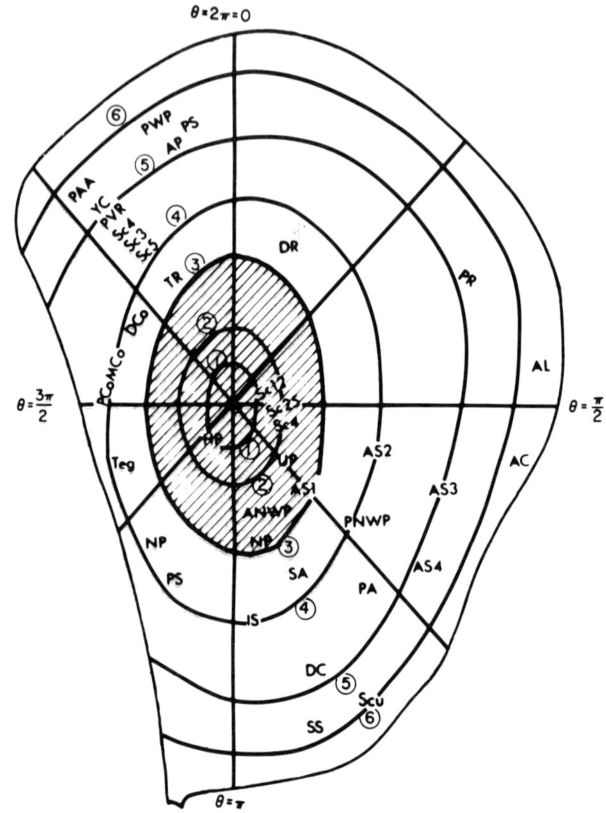

Fig. 8.4. Polar coordinate system on wing disc, with distal wing "high point" region removed. No parts of proximal annulus have missing distal radial values. Therefore, no tissue contact and positional averaging can intercalate the missing distal radial values. See text for discussion.

the <u>convex</u> <u>set</u> of all those positional values derivable by averaging <u>any</u> pairs of positional values present along the cut margin of the proximal annulus.

A more formal definition of a convex set may be given. Consider a two dimensional sheet of cells in which transverse gradients of two scalar properties (concentrations) X and Y exist and supply a unique positional value (X_i; Y_i) to each point in the tissue. Each position in the tissue can be mapped to a unique "image" point in an X, Y "Tissue Specificity Space" [86], yielding an "image" of the tissue in the XY space, Figure 8.5. If the tissue is cut into a fragment which heals and apposes normally nonadjacent tissue regions, diffusive-like averaging will smooth the discontinuities in X and Y in the new cells which form in the wound area. If the original X and Y values are imagined to remain fixed at the two apposed margins, diffusion will yield smooth gradients of X and Y across the new tissue, spanning

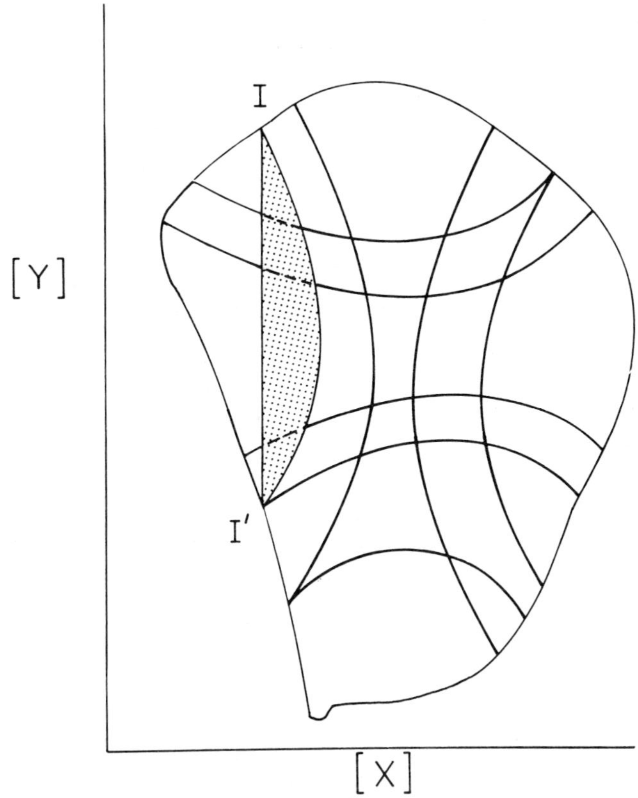

Fig. 8.5. Tissue Specificity Space assigning each point on wing disc a unique value of X and Y. Bowed concave lines are images of straight cuts on the wing disc like those in Figure 8.2. A single cut from I to I' lies along the concave arc from I to I'. Wound healing apposes the wound margin in the posterior fragment. Juxtaposition creates positional discontinuities which are smoothed by diffusion to fill the convex set, shaded area, bounded by the cut margin.

between the bounding values of X and Y in the apposed tissue edges. The image of these intercalated new X and Y values lie on smooth lines in the Tissue Specificity Space, connecting the images of the two apposed old positional values. The collection of such lines, shaded area, Figure 8.5, is the convex set of XY positional values which can be obtained by diffusive smoothing of the juxtaposed wound edges.

The polar coordinate model can be deformed to a cylinder by opening the tissue sheet at the high point origin and mapping it to the top edge of the cylinder, as in a mercator projection. Radial positions become latitude circles along the cylinder. Use of the independent radial and short arc intercalation rules of the polar model allows precise definition of the convex set obtained by apposing any two tissue regions. Removal of the high point region leaves the lower portion of the cylinder corresponding to the proximal annulus. The distal region does not lie in the convex set derivable by intercalation from the proximal annulus.

The implication of this feature of any polar coordinate model is that distal regeneration by a proximal fragment cannot be obtained by diffusive-like averaging of positional discontinuities, and some special rule is needed. In the initial formulation of the polar coordinate model, special rule 3, the complete circle rule was proposed. According to this rule, exposure of a complete circle of angular values at a proximal level leads to regeneration of distal radial values. With the assumption of this rule, the model accounts for the capacity of a truncated amphibian limb to undergo distal transformation, and regenerate distal structures. The same postulate accounts for distal regeneration by a proximal wing disc annulus. Equally, the third rule accounts for duplication of the cultured wing disc high point region from the exposed circle of angular values at a proximal radial level, and similarly explains duplication from the proximal cut surface by a newt hand whose digits are implanted into a host flank. Finally, the third rule accounts for the striking observation that grafting a left hand to a right stump yields two supernumerary limbs at the position of maximal discord of angular values. Such a graft creates two complete circles of angular values at the radial level of the graft. These undergo distal transformation and yield two supernumeraries, with the handedness of the host [31].

The successes of the polar model are considerable, and have led to testing several features of the model. I turn now to an evaluation of weaknesses in this model.

First, it should be stressed that special rule 3, or any modified form of rule 3, is formally equivalent to postulating a mechanism beyond diffusive-like averaging of positional values, to regenerate a missing radial gradient "peak." For example, in the current version of the polar coordinate model ([13], [14]) its authors propose that short arc intercalation itself generates distal regeneration. While the postulate of a special mechanism to regenerate distal radial values is not a flaw, its status should be made explicit. As shown below, distal regeneration does not require special rules in other coordinate systems. If special mechanisms exist to recreate missing gradient values, then those mechanisms are of central importance. They are likely to play a role in the initial establishment of positional gradients as well as subsequent pattern regulation. The forms of special mechanisms which are suggested depend on the choice of coordinate system. Thus, different coordinate systems suggest that different cellular properties beyond diffusive averaging are required to account for the data on pattern formation.

Second, as noted by Winfree [84], in the vicinity of the origin, the gradient of the angular variable in the tissue becomes infinitely steep, since the origin is a singularity in the angular map on the plane. If cell proliferation is proposed to occur proportional to the gradient in radial or angular values, then proliferation near the origin should be unbounded. It is not. As Winfree points out, this implies a lack of independence of radial and angular gradients near the origin.

Third, anterior-posterior and dorsal-ventral developmental axes appear to be established sequentially. This is not deducible, hence not explained, by the proposal that position is measured in polar coordinates. For example, if a radial axis is established first, the amphibian limb bud should be angularly symmetric in transplantation between the time when the first axis and second axis are formed, in contravention to Harrison's classical observation [37]. If angular values are established first, then reversal of anterior-posterior or dorsal-ventral areas should yield similar, not dissimilar results, at each stage, but do not [37].

Fourth, a considerable body of work testing the complete circle rule has shown it to be false in imaginal tissue, cockroach limbs, and amphibian limbs ([40], [43], [48], [70], [73], [77], [81]). It now appears that distal regeneration is roughly proportional to the proximal angular arc present. Confirmation of the complete circle rule would have been striking evidence in favor of a polar model. Its disconfirmation does not rule out such a model, but alternative coordinate systems have the deductive consequence that distal regeneration should be proportional to the proximal arc; hence the data disconfirming the complete circle rule leave a polar model with a postulate as rule 3, while alternatives deduce, hence explain, this phenomenon.

Fifth, accumulating evidence shows that normally duplicating imaginal disc fragments can with variable frequency regenerate some, most, or perhaps all of the complementary domain. Such results are now known in the wing, haltere, leg, and genital discs ([11], [27], [40], [48], [51], [69], [84]). No convincing comparable data are yet available on amphibian or cockroach limbs or other epimorphic systems, although divergent regeneration from symmetrical half limbs in amphibians ([43], [81]) might be due to regeneration by normally duplicating fragments.

The polar coordinate model is constructed such that intercalation (Rules 1 and 2a, b) causes complementary fragments strictly to duplicate and regenerate, Figure 8.3. The model is unable to account for regeneration by normally duplicating disc margin fragments by simple positional averaging. In fact, no proposed coordinate system, coupled with the assumptions of simple diffusive averaging of apposed positional values, can account for extensive regeneration by both the normally duplicating and normally regenerating fragments of a tissue. All coordinate systems share the property that both complementary fragments have identical positional values along their cut margins, hence if both fragments heal in similar ways, apposing similar pairs of positional values, both fragments must reform the same set of structures lying in the convex set bounded by the apposed positional values. Evidence that each complementary disc fragment can regenerate large domains of the other fragment is evidence that mechanisms beyond simple diffusive smoothing must operate during imaginal disc pattern regulation. The polar coordinate model can account for this phenomenon by weakening the short arc intercalation rule to allow long arc intercalation. Formally, this is equivalent to an active mechanism to regenerate angular values not in the convex set reached by diffusive smoothing.

Sixth, mutants in Drosophila are available which can produce completely duplicated legs ([12], [46]). In order for a disc fragment to produce a complete duplicate appendage by apposing cut surfaces of the fragment and positional smoothing, the apposed surfaces would have to form a convex set containing the entire tissue image. Examination of the polar coordinate model, Figure 8.3, shows no single fragment has this property. Within the frame of the polar model, a partial disc fragment can produce a mirror symmetric duplicate, but no fragment of a normal disc can produce both a normal leg, and a complete duplicate appendage.

Finally, the polar model has not yet received a specific molecular or mechanistic interpretation proposing what kinds of variables measure angle or radius, nor how they are established during development.

CARTESIAN COORDINATE SYSTEM

Recently several workers have independently suggested some form of a Cartesian coordinate model to account for the data on epimorphic pattern regulation ([24], [47], [48], [72], [86]). Suggestions by Winfree and myself are nearly identical. Figure 8.6A shows the Drosophila wing disc with roughly orthogonal anterior-posterior and ventral-dorsal gradients of two chemicals, X and Y. Lines of constant concentration are bowed outward on the disc, symmetrically about the "high point." The model assumes that a cut wing disc fragment heals its cut margin, apposing non-adjacent positional values, and that simple diffusion smooths discontinuities in X and Y, and fills in the convex set bounded by the apposed XY pairs along the cut margin. Figure 8.5 shows, for the same Cartesian model the image in Tissue Specificity Space XY, of a straight cut on the disc. As shown in Figures 8.5 and 8.6A the bowing of lines of constant X and Y concentration on the disc implies that a large posterior fragment of a single straight cut will fill the shaded convex set and regenerate anteriorly to the anteriormost value present along its cut margin. The complementary anterior fragment apposes the same pairs of discordant values, and duplicates to the same anteriormost value.

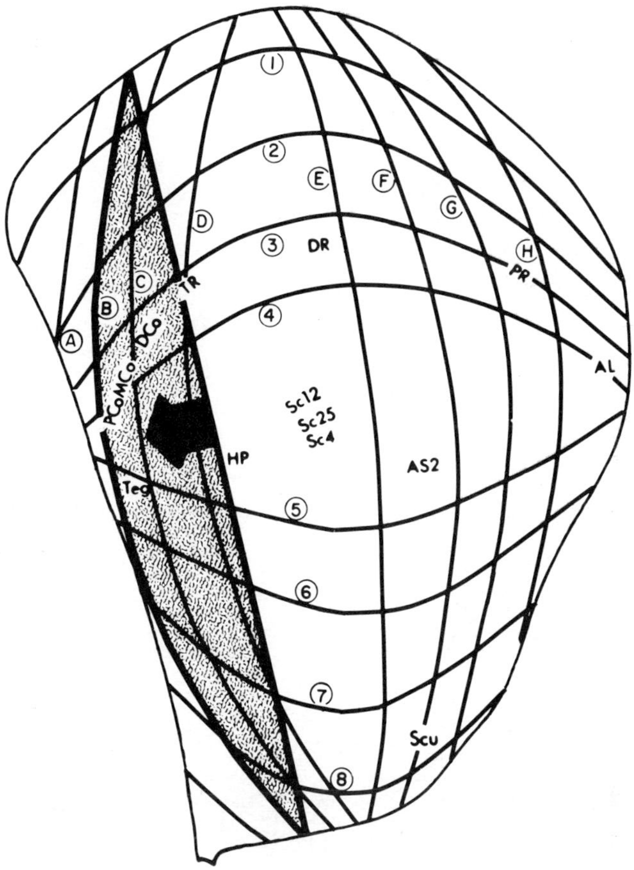

Fig. 8.6A. A Cartesian coordinate system with two monotonic gradients. The "Y" variable has ascending levels 0-9. The "X" variable has ascending levels A-H. A straight cut creates a large posterior fragment which wound heals by folding in half and juxtaposing opposite ends of the cut margin. For this particular cut, this leads to a discontinuity across the wound in Y values, but not in X values. The convexity in lines of constant concentration of each variable, isoconcs, assures that simple diffusive smoothing of X and Y variables yields regeneration out to the anteriormost X isoconc, B, contained along the cut margin of the posterior fragment, stippled area. Similarly, the smaller anterior fragment wound heals the same way and duplicates to the B isoconc. The symmetry of convexity in the X and Y isoconcs on the disc about a central region in the wing disc implies that the polarity of regeneration (heavy arrow) will alter about an apparent "high point" in this region, although cells do not measure position with reference to the high point. Regeneration by a normally duplicating fragment requires an active mechanism beyond simple diffusion to recreate the gradient extremum at the disc margin of the normally regenerating fragment.

Symmetrical convex bowing of X and Y concentrations about the high point insures that the direction of regeneration reverses about the high point. A large anterior fragment from a straight cut will regenerate posteriorly, its posterior complement will duplicate. Similarly, any 1/4 pie wedge fragment which apposes its two cut margins will duplicate, its 3/4 complement will regenerate. Therefore, a Cartesian model yields reversal of the direction of regeneration about the high point, without the assumption that the high point is a locus from which cells measure position.

Distal regeneration is a consequence of simple diffusive smoothing in a Cartesian model. As shown in Figure 8.6B, deletion of the high point region leaves a proximal annulus. Wound healing apposes tissue around the circular cut margin, thereby creating discontinuities in the X and Y gradients. Since the deleted high point region lies in the convex set reached by diffusive smoothing from the proximal annulus, the distal high point region is regenerated. Similarly, the high point region itself purse string closes its circular wound margin, apposing the same pairs of positional valus as did the proximal annulus, and therefore duplicates the X and Y high point values in the new cells of the wound area, Figure 8.6B. The same principle predicts that any proximal arc of tissue will undergo limited distal regeneration of

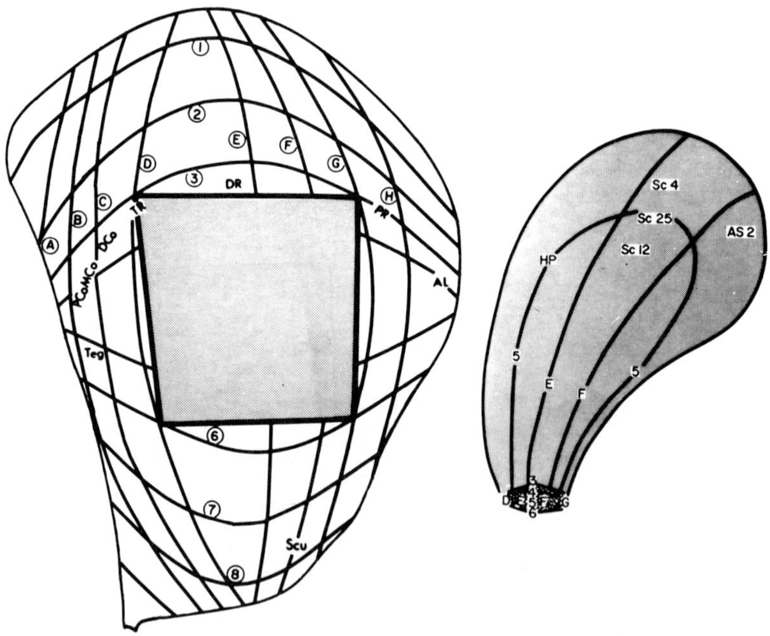

Fig. 8.6B. Removal of a central (distal) region leaves a proximal outer annulus. Wound healing along the cut in the annulus juxtaposes tissues, creating discontinuities in the monotonic X and Y gradients, which are smoothed over by simple diffusive-like processes, leading to distal regeneration by the proximal annulus. The central square containing the high point itself purse string closes juxtaposing tissues that again create X and Y gradient discontinuities across the wound healed zone. Smoothing of these discontinuities causes the positional values in the central square to be duplicated in the new cells of the wound zone, shown in the granular area.

structure lying in the convex set of the positional values along the cut margin, Figure 8.6C. Therefore, limited distal regeneration proportional to the proximal arc present, as recently demonstrated in imaginal discs, and amputated symmetric half limbs of amphibians and cockroaches, is a deductive consequence of a Cartesian model.

A Cartesian model has no singularities with infinitely steep gradients, as does the polar model for angular values near the origin. Therefore, the proposal that cell proliferation is related to gradient can be well believed throughout the domain.

As shown in Figure 8.7, the Cartesian model explains the striking observations that grafts of distal left limbs to proximal right stumps can generate two supernumerary limbs with the handedness of the host. This property now appears to be a coordinate free consequence of local coordinates in the plane, and the postulate of

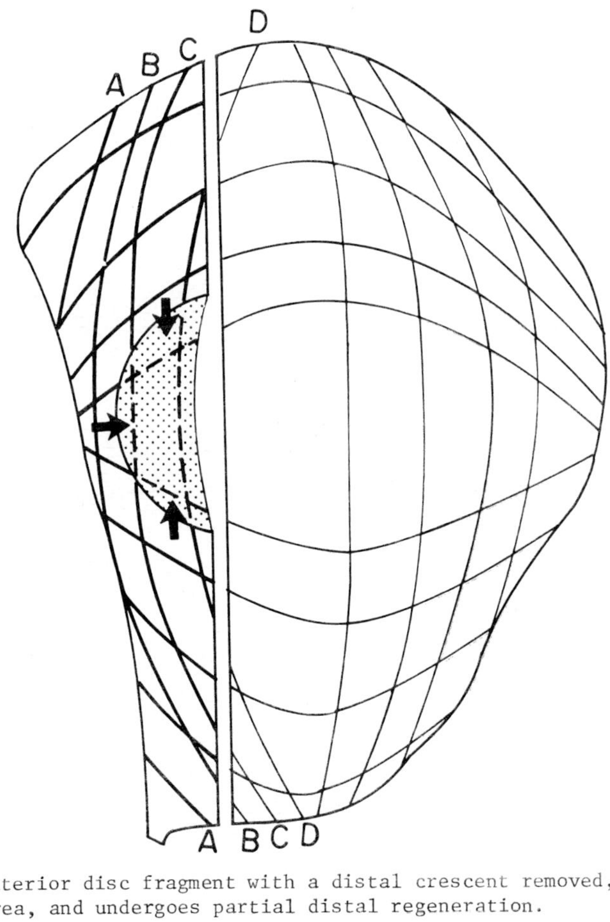

Fig. 8.6C. Anterior disc fragment with a distal crescent removed, fills convex set, shaded area, and undergoes partial distal regeneration.

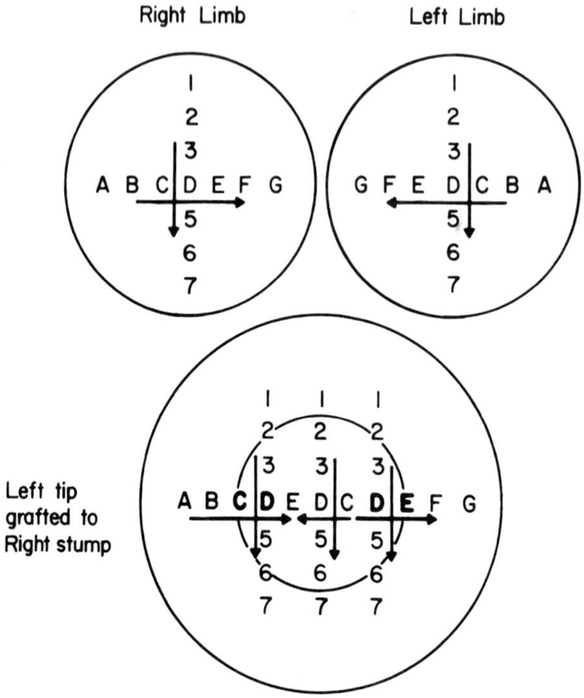

Fig. 8.7. Schematic picture of anterior-posterior and dorsal-ventral Cartesian X, Y coordinates for left and right limbs, projected onto the plane. The position at the mid values of the two variables, 4, D, corresponds to the distal limb tip. Grafting a left distal limb to a right proximal stump leads to smoothing of discontinuities in X and Y (bold letters, CD and DE) and the formation of two supernumerary distal limb tips with the handedness of the host proximal stump, at the positions of maximal disparity of host and graft axes.

intercalary smoothing. With minor distortion from symmetry, the Cartesian model also has the consequence that 180° graft rotations of the left hand onto the left stump can yield variable numbers of supernumeraries, including two supernumerary limbs of opposite handedness.

Finally, unlike the polar model, a Cartesian model accords directly with the widespread data suggesting that anterior-posterior axes are established prior to dorsal-ventral axes.

Any Cartesian model suffers certain defects analogous to those of the polar model. The "high point" region in the polar model lies outside the convex set of the proximal annulus. Its regeneration requires special mechanisms. In a Cartesian model, the extreme disc margins lie outside the convex sets of their complements, Figure 8.5. Thus, as shown in Figure 8.6A, the large posterior fragment only regenerates partially. It does not reach the anterior disc margin. Similarly, the narrow anterior fragment duplicates partially. Either extreme bowing of lines of constant concentration or special mechanisms to recreate X and Y gradient maxima and minima are required to account for the observed capacity of disc fragments to regenerate or duplicate completely.

Like the polar model, no Cartesian model can explain the capacity of normally duplicating fragments to regenerate large domains of the complementary disc fragment by diffusive smoothing. Nor can a Cartesian model account by diffusive smoothing alone, for mutants which form complete limbs and complete duplicate limbs, since no subregion of a Cartesian coordinated tissue can have the entire tissue image in its convex set. The assumption of a Cartesian model leads to the suggestion that cells in a tissue have special mechanisms capable of reforming X and Y gradient maxima and minima lying outside the convex sets of cultured fragments. Explicit mechanisms are described below.

HIGHER DIMENSIONAL MODELS: COMPACTNESS

I describe only briefly the general problem associated with the assumption that 3 or more chemical gradients in a two dimensional tissue sheet supply positional information. Consider a three variable model, with a Tissue Specificity Space XYZ. The image of the tissue is a two dimensional surface embedded in the 3 chemical Tissue Specificity Space. Suppose that the image surface is curved, not planar. Then, due to the curvature, apposition of non-adjacent tissue regions can result in intercalation of positional values lying outside of the two dimensional image of the tissue in the XYZ space. Thus a higher dimensional model typically confronts the problem of "compactness." Are all positional values lying within the convex set of any two tissue fragments themselves "meaningful"--i.e., part of the tissue image. The 3 variable model can meet the problem of curvature of the tissue image in two ways: 1) by assuming processes beyond diffusive smoothing which constrain the intercalated values back into the curved image; 2) by assuming lines of redundant "equivalent" positional information in X, Y, Z space, each transversal to the image and piercing it at a single spot. Transversal lines to the tissue image can fill 3 space in such a way that intercalation between normally non-adjacent wound edges leads to new positional values lying in the convex set bounded by the apposed values, which are "meaningful." The Spherical Coordinate Model discussed next uses one form of the latter solution.

SPHERICAL COORDINATE MODEL

Russell [66] proposed a spherical coordinate model using three orthogonal X, Y and Z gradients to form a Tissue Specificity Space. Position in a tissue is specified by a solid angle, θ, ϕ corresponding to latitude and longitude angles. Each positional value is a unique ray at a constant X:Y:Z ratio emanating from the origin. The longitudinal angle ϕ, is defined by the ratios of X and Y in the equatorial plane, and the latitude angle θ, is defined by the X:Z, or Y:Z ratio. The image in XYZ space of a two dimensional tissue is a two dimensional surface pierced by a set of solid angle rays emanating from the origin, which itself does not normally lie in the physical tissue.

Russell's model is elegant in a number of respects. The rays are each lines of equivalent positional value in XYZ space, thus, the image is compact. Any two non-adjacent fragments which are apposed will fill their convex set by simple diffusion, and the resulting values will be a surface in XYZ space pierced by rays from the origin, hence "meaningful." Consequently, the spherical model can account for almost all the available data using only the concept of diffusive smoothing. This is most easily visualized in the wing disc by remembering that the disc is topologically a spherical surface, the disc proper, backed by the peripodal membrane. Let this tissue spherical surface be embedded in XYZ space such that each solid angle pierces the closed two dimensional surface of the image once. Therefore the image surrounds the origin. If a narrow anterior fragment of the disc is cut, the margin purse string wound heals, and smoothing of X, Y, and Z discontinuities in the new cells forms a second surface in XYZ space whose edges join those of the original anterior fragment image, Figure 8.8. Rays from the origin pierce both the new surface and the original anterior image surface, Figure 8.8, hence the narrow fragment duplicates completely. The large posterior fragment wound heals similarly, and forms the same new image surface in XYZ space in the new cells of the wound area. These are pierced by the same rays that pierce the anterior fragment, hence the posterior fragment regenerates completely.

The model is spherically symmetric. Therefore not only do narrow anterior, posterior, dorsal or ventral fragments duplicate while their large complements regenerate, but a distal fragment containing the high point region will purse string close, creating a second image surface in the new cells in the wound area. This second surface will be pierced by the same rays that pierce the high point region, therefore it will duplicate the high point. Equally the proximal annulus will regenerate distally. Finally, this model directly explains distal regeneration proportional to the proximal arc cultured, and the incidence of supernumerary limbs.

Despite its strengths, the spherical model suffers some weaknesses. First, it requires cells to assess position by measuring the ratio of two or three scalar variables over a continuous range of their concentrations with respect to a reference concentration of each at the origin. Plausible mechanisms to do so need to be brought forward. Second, the spherical model, like the polar model, has a

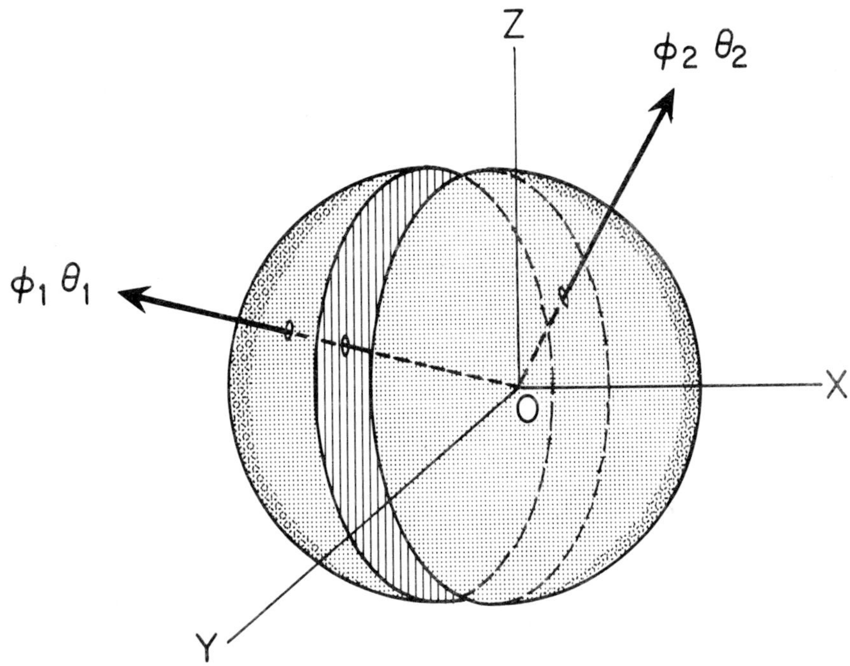

Fig. 8.8. Russell's [66] spherical coordinate model. X, Y, and Z are orthogonal gradients. Position is specified by ratio of X:Y:Z: with respect to origin, 0; each ratio is a ray at a unique solid angle, ϕ, θ. Figure shows disc as a spherical surface (dotted region), with an anterior fragment cut off. Wound healing in anterior fragment creates an additional new XYZ surface (parallel lines). Rays $\phi_1\theta_1$ pierce both surfaces, hence anterior fragment duplicates. Posterior fragment forms same new XYZ surface and regenerates.

singularity at the north pole high point, where all longitudinal angular values meet. The hypothesis that positional gradients regulate proliferation confronts infinitely steep longitudinal angular gradients at the "high point."

Finally, while the spherical model is very appealing, it is unable to account for all the data by simple diffusive averaging of gradient discontinuities. Like the polar and Cartesian models, diffusion in the spherical model can yield only duplication and regeneration by complementary fragments. It cannot account for regeneration by normally duplicating fragments, nor the existence of mutants which form normal limbs, and complete duplicate limbs. To account for these phenomena, it is necessary to imagine cellular processes capable of recreating missing X, Y, or Z maxima and minima not in the convex set of the cultured fragment.

SEQUENTIAL GENERATION OF POSITIONAL AXES

Review of the polar, Cartesian, and spherical coordinate models demonstrates that the single property of intercalation of intervening positional values is very powerful, yet not adequate to account for regeneration by normally duplicating fragments, nor for the capacity of some mutants to produce complete normal plus duplicate appendages. In the Cartesian and Spherical models, mechanisms are needed which can recreate gradient maxima and minima. One interpretation of the polar model (*) has the same requirements. In the present section I discuss the role reaction-diffusion instabilities may have in the establishment and maintenance of positional gradients during development.

It is now well known that model biochemical systems in which synthesis and degradation of chemical species are coupled, and their diffusive transport occurs, can lead to the formation of spatially inhomogeneous gradients of the constituents in the domain ([4], [34], [35], [41], [49], [53], [58], [83]). As a particular model, we postulate a single biochemical system of two components, with concentrations X(r,t) and Y(r,t) at position r, time t, which are being synthesized and destroyed at rates F(X,Y) and G(X,Y) and are diffusing throughout a tissue. The equations for this system are:

$$\frac{\partial X}{\partial t} = F(X,Y) + D_1\nabla^2 X$$
$$\frac{\partial Y}{\partial t} = G(X,Y) + D_2\nabla^2 Y \qquad (8.1)$$

which are chosen to have a spatially homogeneous steady state Xo, Yo at which F(Xo, Yo) = G(Xo, Yo) = 0.

Analysis of System (8.1) begins by linearizing about the spatially homogeneous steady state by substituting X(r,t) = Xo + x(r,t), Y(r,t) = Yo + y(r,t) and retaining only terms up to first orders in x and y in a Taylor expansion of F(X,Y) and G(X,Y). The resulting linear equations in deviations x and y from the homogeneous state are:

$$\frac{dx}{dt} = K_{11}x + K_{12}y + D_1\nabla^2 x$$
$$\frac{dy}{dt} = K_{21}x + K_{22}y + D_2\nabla^2 y \qquad (8.2)$$

The stability of the spatially homogeneous steady state of the linearized equations to spatially inhomogeneous perturbations in the concentration of x or y may be analyzed by evaluating the determinant of the matrix

$$\begin{vmatrix} K_{11} - k_i^2 D_1 - \lambda & K_{12} \\ K_{21} & K_{22} - k_i^2 D_2 - \lambda \end{vmatrix} = 0 \tag{8.3}$$

yielding the dispersion relation between the two temporal eigenvalues λ_{i+}, λ_{i-}, and the spatial eigenvalue $k^2 i$.

$$\lambda_i \pm = \frac{1}{2}[K_{11} + K_{22} - k^2(D_1 + D_2)] \pm \tag{8.4}$$

$$\frac{1}{2}\sqrt{[K_{11} + K_{22} - k^2(D_1 + D_2)]^2 - 4[k^4 D_1 D_2 - k^2(D_2 K_{11} + D_1 K_{22}) + K_{11}K_{22} - K_{21}K_{12}]}$$

Here $k_i = \sqrt[2]{k_1^2}$ is inversely proportional to the wavelength of the perturbation ℓ_i. For a given wavelength perturbation ℓ_i, corresponding to a specific value of k_i, if both λ_{i+} and λ_{i-} have negative real parts, that perturbation decays. In an unbounded domain, those perturbations with wavelengths corresponding to values of k_i for which the dispersion relation has at least one temporal eigenvalue with positive real part, will grow in amplitude in time and create a spatial pattern.

With appropriate constraints on the linearized reaction and diffusion constants ([35], [49]), the dispersion relation $\lambda = f(k)$ is positive for a restricted range of k, $k_1 > k_i > k_2$, hence for a restricted range of wavelength ℓ, $\ell_1 < \ell_i < \ell_2$. In this case, System (8.1) acts as an amplifier, selecting and amplifying from thermal noise those wavelength components ℓ in the range between ℓ_1 and ℓ_2.

If System (8.1) exists in a bounded spatial domain, and no flux boundary conditions are imposed, then a chemical pattern will be established only if it simultaneously has a wavelength in the range $\ell_1 - \ell_2$, and also satisfies the boundary condition that the gradient have zero component normal to the boundaries. Assuming no flux boundary conditions, the shapes of chemical patterns which can "fit" on to the spatial domain must satisfy the Laplacian diffusion operator, and are an infinite series of specific eigenfunction patterns determined by the geometry of the domain. The first several eigenfunctions on a wing disc shape are shown in Figure 8.9A-D [18].

ESTABLISHMENT OF A FIRST POSITIONAL AXIS

As the spatial domain in which System (8.1) acts grows gradually larger, a succession of the discrete eigenfunctions or their superposition will sequentially become established in the domain [49]. For any reaction-diffusion system which acts as an amplifier, perturbations below some minimal wavelength, ℓ_1, will die away. Therefore, if the maximum physical length of the tissue is less than ℓ_1, the spatially homogeneous state is stable. As the tissue grows in length past ℓ_1, the first eigenfunction of the tissue domain will bifurcate from the homogeneous solution.

Fig. 8.9A.

Fig. 8.9B.

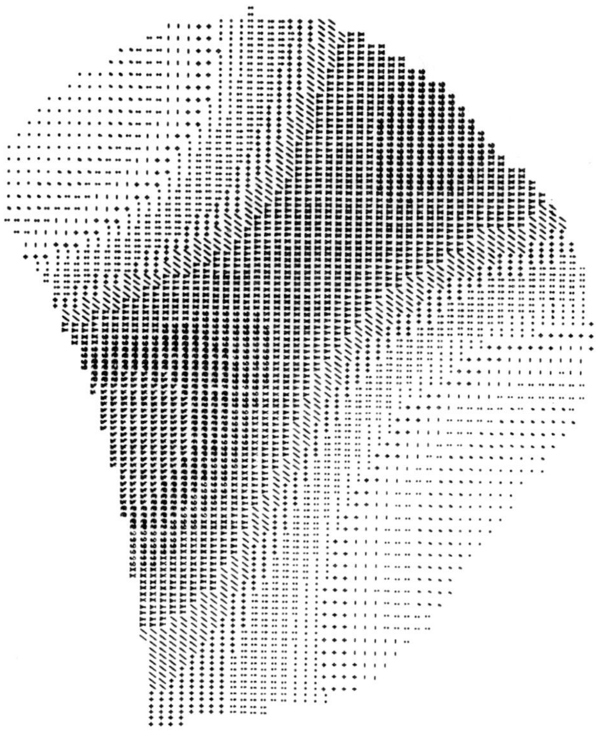

Fig. 8.9C.

A fundamental theorem of bifurcation theory guarantees that the first pattern will be a stable solution of the full non-linear equations [68]. Therefore, as a wing disc shape gradually enlarges, no chemical pattern forms for lengths below ℓ_1, then eigenfunction 1, Figure 8.9A, grows out of the homogeneous state, and forms a monotonic gradient along the long axis of the tissue.

It has been recognized for some time that this primary bifurcation provides a means to establish a first dimensional axis of positional information in a growing domain. The new question I wish to address is: "How can a reaction-diffusion system like (8.1) establish two transverse axes in a growing tissue?"

In the linearized equations of (8.1) or any similar system, each eigenfunction pattern which forms in the tissue is a spatially inhomogeneous distribution of a fixed ratio of the underlying chemical variables x and y, measured as deviations from the homogeneous steady state Xo,Yo. This property follows from the fact that the unstable wavelength ℓ_i which is amplifying on the tissue, corresponds to a unique scaling factor k_i. Each unique choice of k_i specifies a particular set of values in Matrix (8.3) having at most one positive temporal eigenvalue ℓ_i; and a

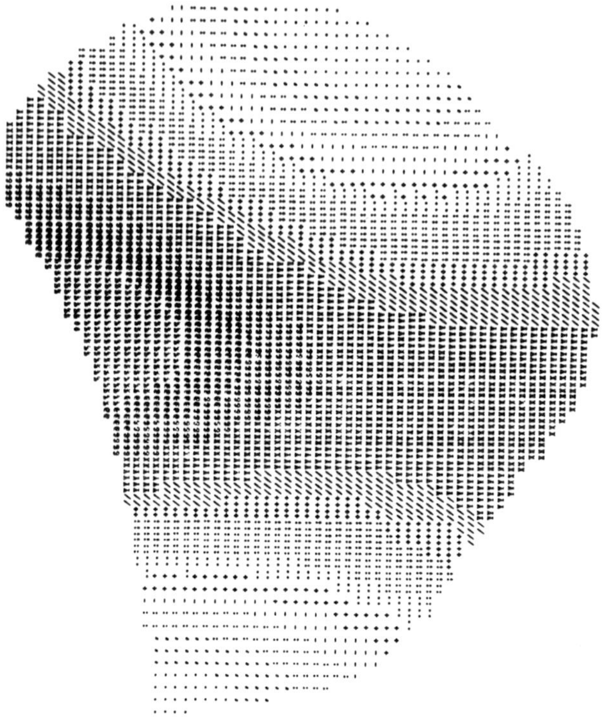

Fig. 8.9D.

Fig. 8.9A, B, C, D. First four eigenfunction patterns on wing disc as it enlarges. Concentration coded by visual density (from Bunow et al. [18]).

unique corresponding eigenvector $x_i:y_i$, which is a fixed ratio of the linearized deviations of x and y from Xo,Yo. This constraint implies that lines of constant concentration of X on the tissue parallel those of Y. Therefore although two chemical gradients are present in a two dimensional domain, they supply at most one dimension of positional information. More generally, if N chemical species are in System (8.1) $N \geq 2$, all N occur as a fixed ratio when a single temporal eigenvalue λ_i is positive, hence lines of constant concentration of all N are parallel to one another and supply only one positional axis. Further, if any single eigenfunction pattern, e.g., Figure 8.9C, is amplified alone in the tissue through a single positive eigenvalue, the ratio of all N species is again fixed, and at most one positional gradient is established.

ESTABLISHMENT OF A SECOND POSITIONAL AXIS

It has not been recognized that the familiar class of 2 variable reaction-diffusion systems capable of establishing a first axis can sequentially establish transverse axes in growing two dimensional domains. This can be exemplified on a wing disc shape. If the range of wavelengths ℓ_1 and ℓ_2 which are amplified by the linearized System (8.1) is made broad enough, then as the tissue domain enlarges past ℓ_1, the first eigenfunction mode, Figure 8.9A, forms. As the two dimensional domain enlarges further, both the first ventral dorsal monotonic pattern and the second anterior-posterior monotonic eigenfunction pattern Figure 8.9B, satisfy the boundary conditions and may be amplified together. If the length and width of the tissue are unequal, each mode corresponds to a different specific wavelength ℓ_i, ℓ_j ($\ell_i > \ell_j$), and the corresponding scaling factors k^2 are unequal, $k^2 i = k^2 j$. Therefore, the eigenvector ratio of the chemical variables x and y, derived from Matrix (8.3) differs for the two modes. The superposition of the two distinct modes, each amplifying a different eigenvector, yields non-parallel (transverse) gradients of X and Y in the tissue. As a second example, on a growing rectangle whose sides are of unequal length, $L_1 > L_2$, the linearized equations sequentially amplify the first eigenfunction creating parallel X and Y monotonic gradients in the L_1 direction, then amplify the superposition of the first two monotonic eigenfunctions in the L_1 and L_2 directions as the rectangle enlarges to an appropriate size creating transverse gradients of X and Y:

$$c_1\begin{bmatrix} x_1 \\ y_1 \end{bmatrix} e^{\lambda_1 T} \cdot \cos \frac{\pi r_1}{L_1}\begin{bmatrix} x_1 \\ y_1 \end{bmatrix} + c_2\begin{bmatrix} x_2 \\ y_2 \end{bmatrix} e^{\lambda_2 T} \cdot \cos \frac{\pi r_2}{L_2}\begin{bmatrix} x_2 \\ y_2 \end{bmatrix} \tag{8.5}$$

The first major property of this class of models, therefore, is the prediction that transverse positional axes should form <u>sequentially</u> as the tissue grows in size.

The linearized analysis of (8.5) is of limited value for several reasons. The amplitude of each of the two superimposed modes grow exponentially without bound. Whichever mode grows faster (λ_1 or λ_2) eventually dominates the superposition, therefore, a single eigenvector $[x_1:y_1]$ or $[x_2:y_2]$ dominates the pattern and the underlying X and Y gradients become parallel. Second, the contributions of the first and second patterns to the superposition are given by arbitrary initial conditions, c_1, c_2. Finally, each eigenfunction pattern can occur in two opposite forms or branches, with maxima replacing minima. Which is established depends entirely on initial conditions. The conjunction of these defects implies that the relative orientation of X and Y gradients on the tissue is not uniquely fixed by tissue geometry in a linear theory.

These defects of a linear model are fully met by a fairly broad class of nonlinear reaction-diffusion systems in an asymmetric two dimensional tissue such as the wing disc. A number of workers have proposed nonlinear models which are known to bound the amplitudes of chemical patterns in one spatial dimension ([4], [41], [52], [53]) and two dimensions ([18], [49]). Generally, the establishment of the first pattern parallels the linear analysis, except that for large amplitudes, the pattern may deviate from the eigenfunction pattern. As the spatial domain enlarges, the second pattern commonly is gradually established through a secondary bifurcation from the primary spatially inhomogeneous monotonic pattern, yielding at each tissue size a fixed mixture or amplitude of distorted first and second modes ([2], [42]). Finally, in a spatially asymmetric two dimensional domain, the cube of the eigenfunction is almost always nonzero. From bifurcation theory [68], this implies that, as the tissue enlarges, only one of the two branches (or forms) of the primary monotonic pattern emerges smoothly from the homogeneous state, while the other branch is stable but requires a macroscopic perturbation to be attained. For at least some nonlinear systems, only one branch of the subsequent secondary bifurcation emerges smoothly from that first mode ([2], [3], [50]). The novel implication is that the entire sequential establishment of transverse gradients in two dimensional domains, their shapes and orientations, can be uniquely determined by the changing tissue size and geometry, and the kinetic parameters of the non-linear model.

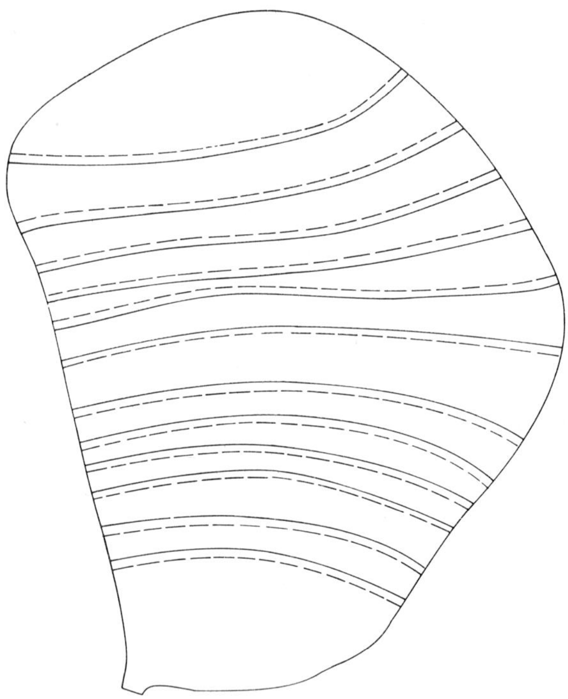

Fig. 8.10A. Establishment of a First Axis. First longitudinal monotonic spatially inhomogeneous steady state of nonlinear system (8.6), corresponding to first eigenfunction, Figure 8.9A. Lines of constant X concentrations (solid lines) parallel lines of constant Y concentrations (dashed lines). Spacing between lines of constant X concentrations, or of Y concentrations are not strictly proportional to gradient slope, but are used to indicate shapes of X and Y gradients.

In Figure 8.10A-B, I show the results of simulations of one non-linear model given by:

$$\frac{\partial X}{\partial t} = -AX + \frac{BY^6}{1 + Y^6} + D_1\nabla^2 X$$
$$\frac{\partial Y}{\partial t} = -CX + \frac{D(Y^6 + b)}{1 + Y^6} + D_2\nabla^2 Y \qquad (8.6)$$

with parameter values specified in the legend. As a wing disc shape enlarges, a first (Figure 8.9A) ventral-dorsal monotonic gradient of X and parallel gradient of Y is established. As the disc enlarges further and the underlying anterior-posterior 2nd mode (Figure 8.9B) begins to contribute to the pattern, the X and Y gradients

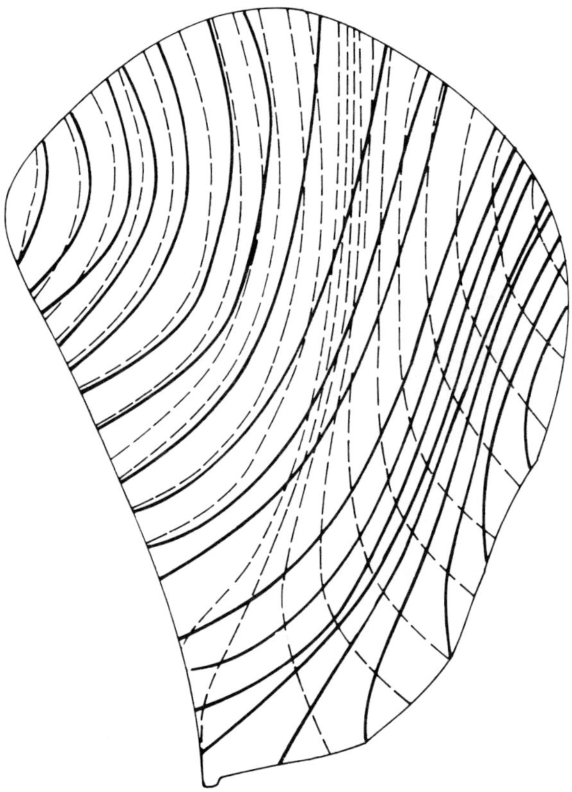

Fig. 8.10B. Establishment of Second Positional Axes. As wing disc grows, second eigenfunction, 8.9B, makes a contribution to mixed nonlinear steady state solution. X and Y gradients rotate away from positions in 8.10A, and lines of constant X concentrations (solid lines) separate from lines of constant Y concentration (dashed lines) creating transverse gradients and a second positional axis. See text. Parameter values used for the nonlinear model (8.6): A = 7.8; B = 15.6; C = 1; D = 1.87; b = .2; D_1 = 16; D_2 = 1.

both rotate slowly on the disc, Y faster than X, gradually establishing distinct nonparallel transverse, but not orthogonal, monotonic X and Y gradients spanning the tissue, Figure 8.10B. The model sequentially establishes transverse axes and two dimensions of positional values through two gradients whose forms and orientations are initially identical, but gradually come to differ as the tissue enlarges. The positional field is uniquely determined by the shape trajectory of the tissue, and the specific non-linear model. Several mathematical features of this class of models are now being explored.

NOVEL PREDICTIONS

The inherent simplicity and minimality of this reaction-diffusion mechanism commends it as a serious model for the establishment of transverse positional axes and positional information in developmental fields. A more complex three variable model can yield three transverse gradients which might underly Russell's spherical coordinate model. Yet more complex models with two independent two variable subsystems can generate transverse gradients. The minimal model has a number of predictive consequences.

1. Positional axes are established sequentially, as seen in several systems.

2. The longer axis in a field is established before the shorter. This expectation appears to be correct at least in the case of primary axes in insect embryos [67].

3. During establishment of the second axis, the positional gradients may rotate on the tissue. Evidence consistent with this prediction has been described by Duranceau [27] for the developing wing disc. However, uncertainty about "equivalent cuts" on discs at different stages precludes clear interpretation of her results.

4. The axes are transverse, but oblique, not orthogonal.

5. Since the monotonic gradients have values above and below the homogeneous steady state values XoYo, that pair of values will typically occur at least once at an interior point in the tissue. The model predicts that very small tissue fragments cannot maintain gradients and will decay toward the steady state, therefore small fragments should drift toward the map position in the tissue corresponding to the steady state. Strub [78] and Krivi and Schneiderman [51] have found evidence that dissociated and reaggregated proximal leg or wing imaginal fragments from small regions transform distally. Other interpretations of these results are possible but "drift" toward a distal "steady state" is, at present, consistent with the results.

Predictions of the two variable nonlinear reaction-diffusion model with regard to duplication and regeneration are a combination of properties arising from simple passive diffusive averaging of positional discontinuities, and properties arising from the active dynamics and boundary conditions.

7. The X and Y transverse gradients are monotonic across the tissue. If the distal "high point" region, or any interior circle, is removed, the proximal annulus should regenerate the distal region by simple diffusive intercalation, while the distal fragment should purse string heal and duplicate. Further, any 1/4 pie wedge piece should duplicate, its 3/4 complement should regenerate by simple intercalation.

8. Lines of constant X and Y concentration are not bowed symmetrically about an interior "high point" but are determined by the eigenfunctions of the tissue shape. Therefore a narrow wing disc margin fragment and its large complement from straight cuts may not duplicate and regenerate in the observed directions by simple diffusive smoothing after apposing and healing the cut surface.

The two variable reaction-diffusion model explains the occurrence of duplication and regeneration by such small and large fragments of the tissue on a novel basis. The first two mixed modes are each monotonic eigenfunctions with wave number 1, Figure 8.9A, 8.9B. The two next eigenfunctions are symmetric patterns of wave number 2, having one maximum and two minima, Figures 8.9C and 8.9D. Recently Hiernaux and Erneux [42] showed in a one dimensional domain that two variable non-linear reaction-diffusion systems could admit the coexistence of both a monotonic gradient, and the first symmetric gradient as stable alternative solutions of the equations. Cutting of the monotonic gradient into a large and small fragment lead to the reformation of the entire monotonic gradient in the large fragment, and the assumption of the symmetrical mode in the small fragment as it grows in length, yielding its duplication.

It is not yet established, but likely that a two variable reaction-diffusion system in an asymmetric two dimensional spatial domain can admit both a mixture of the first two monotonic gradients and a coexisting mixture of the first two symmetric modes as stable alternative solutions of the equations. This model will yield duplication of a fragment, not only by intercalation of the convex set obtained by diffusive averaging, but by a macroscopic transition to a symmetric chemical mode. Thus, this class of models predicts that wound healing of apposing edges and averaging of discordant positional values may not be necessary for duplication, since other perturbations can induce transition from the monotonic to the coexisting symmetric solution. Parallel suggestions to account for homeotic mutants forming mirror symmetric double abdomen in Drosophila such as bicaudal have been made by Meinhardt [53] and Deak [25].

9. The coexistence of monotonic and symmetric solutions of the equations has the further consequence that the smaller tissue fragment, during growth can regenerate the entire monotonic gradient rather than switch to the symmetric mode. Thus both large and small fragments can regenerate the entire pattern. This enables the model to account for regeneration by normally duplicating imaginal fragments ([11], [27], [40], [48]) and divergent distal regeneration in amphibian symmetrical half limbs [43].

10. If monotonic and symmetric solutions coexist as stable solutions in cultured tissue fragments, the fragment may either duplicate or regenerate, or do mixtures of both, as recently observed in cultured wing disc fragments [48].

11. Finally, as the spatial domain grows, establishment of the symmetric modes can yield mirror duplicate patterns having the full range of X and Y values. Thus, the model can explain the occurrence of mutants such as those in Drosophila which cause formation of both a normal leg and a complete duplicate leg ([12], [46]), which is inexplicable by positional smoothing alone.

These predictions have not been systematically tested. One aim of this review is to stimulate experiments further testing the polar, Cartesian and spherical models, and the postulated reaction-diffusion systems which might underlie those alternative positional coordinate systems. This discussion has focused on epimorphic pattern regulation, but it is clear that the assumption of reaction-diffusion systems which recreate gradients assimilates epimorphic and morpholactic pattern regulation into a unified spectrum of models governed by the extent to which tissue enlargement is required for reestablishment of gradient maxima and minima. I have not discussed a number of important problems such as the apparent coexistence of segmentally repetitive and whole limb gradients [72].

SUMMARY

This article has reviewed and compared the polar, Cartesian and spherical coordinate models for epimorphic pattern regulation, focusing largely on imaginal tissues of Drosophila. Geometric features of the three models lead to variable strengths and weaknesses and differing properties by which they can be experimentally distinguished. Two or three variable reaction-diffusion systems which establish mixtures of two eigenfunctions on the tissue are an important class of models capable of accounting for major features of the establishment and reestablishment of positional information in developmental fields.

*One molecular interpretation of the polar model would propose three orthogonal gradients X, Y, Z. The ratio of X and Y with respect to reference X and Y values on the Z axis would define the angular measure, modulo 2π, as in the spherical model. The concentration of Z, whose maximum would be the "high point," would define radial distance. Thus planes of constant [Z] in XYZ space define radial distance from the high point, and rays of constant X:Y ratio in each plane, define angular measure at that radial level. Short arc angular intercalation emerges as diffusive smoothing of X and Y discontinues, radial intercalation as smoothing of Z discontinues. Active distal regeneration is the need to recreate a Z maximum, and active long arc regeneration is the need to reform missing X and Y maxima or minima not in the convex set of the cultured fragments. This interpretation makes the polar model similar to the Cartesian and Spherical models, and suggests that reaction-diffusion mechanisms might be used to establish any of the three coordinate systems.

ACKNOWLEDGMENTS

It is a pleasure to thank Drs. Russell, Winfree, Hunt, Kernevez, Herschkowitz-Kaufman and Auchmuty, and M. Weir for many stimulating discussions of these problems. This work was partially supported by grants ACS CD-30 and NIH GM22341.

REFERENCES:

1. Apter, M. J., Cybernetics and Development, International Series of Monographs in Pure and Applied Biology/Zoology Division, Vol. 29, Pergamon Press, New York, London, 1966.

2. Auchmuty, J. F. G., Qualitative effects of diffusion in chemical systems. Lectures on Mathematics in the Life Sciences 10: 49 (1977).

3. Auchmuty, J. F. G. and Nicolis, G., Bifurcation analysis of nonlinear reaction-diffusion equations. I. Evolution and the steady state solutions, Bull. Math. Biol. 37: 323 (1975).

4. Babloyantz, A. and Hiernaux, J., Models for cell differentiation and generation of polarity in diffusion-governed morphogenetic fields, Bull. Math. Biol. 37: 637 (1975).

5. Bohn, H., Analyse der Regenerationsfahigkeit der Insektenextremität durch Amputations und Transplantationsversuche an Larven der Afrikanischen Schabe Leucophaea maderae Fabr. (Blattaria). I. Regenerationspotenzen, Wilhelm Roux's Archives 156: 49 (1965).

6. Bohn, H., Interkalare Regeneration und segmentale Gradienten bei den Extremitäten von Leucophaea-Larven (Blattaria), I. Femur und Tibia, Wilhelm Roux's Archives 165: 303 (1970).

7. Bohn, H., Interkalare Regeneration und segmentale Gradienten bei den Extermitäten von Leucophaea-Larven (Blattaria). III. Die Herkunft des Interkalaren Regenerates, Wilhelm Roux's Archives 167: 209 (1971).

8. Bohn, H., The origin of the epidermis in the supernumerary regererates of triple legs in cockroaches (Blattaria), J. Embryol. Exp. Morphol. 28: 85 (1972).

9. Bryant, P. J., Pattern formation in the imaginal wing disc of Drosophila melanogaster: fate map, regeneration and duplication, J. Exp. Biol. 193: 49 (1975).

10. Bryant, P. J., Bryant, S. V. and French, V., Biological regeneration and pattern formation, Sci. Amer. 237: 66 (1977).

11. Bryant, P. J. and Hsei, B. W., Pattern formation in asymmetrical imaginal discs of Drosophila melanogaster, Amer. Zool. 17: 595 (1977).

12. Bryant, P. J. and Schubiger, G., Giant and duplicated imaginal discs in a new lethal mutant of Drosophila melanogaster, Dev. Biol. 24: 233 (1971).

13. Bryant, S. V., Pattern regulation and cell commitment in amphibian limbs, in: The Clonal Basis of Development, 36th Symp. Soc. Develop. Biol., p. 63, 1978.

14. Bryant, S. V., French, V. and Bryant, P., Distal regeneration and symmetry, submitted for publication.

15. Bryant, S. V. and Iten, L. E., Supernumerary limbs in amphibians: Experimental production in Notophthalmus viridescens and a new interpretation of their formation, Dev. Biol. 50: 212 (1976).

16. Bulliere, D., Interprétation des régénerats multiples chez les insectes, J. Embryol. Exp. Morphol. 23: 337 (1970).

17. Bulliere, D., Utilisation de la régéneration intercalaire pour l'étude de la détermination cellulaire au cours de la morphogenèse chez Blabera craniifer (Insecte Dictyoptere), Dev. Biol. 25: 672 (1971).

18. Bunow, B., Kernevez, J.-P., Joly, G. and Thomas, D., Pattern formation by reaction-diffusion instabilities: Applications to morphogenesis in Drosophila, J. Theor. Biol. 84: 629 (1980).

19. Butler, E. G., Regeneration of the urodele forelimb after reversal of its proximo-distal axis, J. Morphol. 96: 265 (1955).

20. Carlson, B. M., Civiletto, S. E. and Goshgarian, H. G., Nerve interactions and regenerative processes occurring in newt limbs fused end-to-end, Dev. Biol. 37: 248 (1974).

21. Chung and Cooke, Polarity of structure and of ordered nerve connections in the developing amphibian brain, Nature 258: 126 (1975).

22. Cohen, N., Amphibian transplantation reactions: A review, Amer. Zool. 11: 193.

23. Cooke, J., The emergence and regulation of spatial organization in early animal development, Annu. Rev. Biophys. Bioeng. 4: 185 (1975).

24. Cummins, F. W. and Prothero, J. W., A model of pattern formation in multicellular organisms, Collective Phenomena 3: 41 (1978).

25. Deak, I. I., A model linking segmentation, compartmentalization and regeneration in Drosophila development, J. Theor. Biol. 84: 477 (1980).

26. Dent, J. N., A study of regenerates emanating from limb transplants with reversed proximodistal polarity in the adult newt, Anat. Rec. 118: 841 (1954).

27. Duranceau, C., Control of growth and pattern formation in the imaginal wing disc of Drosophila melanogaster, Ph.D. thesis, Univ. Calif. Irvine, Irvine, Calif., 1977.

28. Erneux, T. and Hiernaux, J., Transition from polar to duplicate patterns, J. Math. Biol. 9: 193 (1980).

29. French, V., Leg regeneration in the cockroach, Blattella germanica. I. Regeneration from a congruent tibial graft-host junction, Wilhelm Roux's Archives 179: 57 (1976).

30. French, V., Leg regeneration in the cockroach, Blattella germanica. II. Regeneration from a noncongruent tibial graft/host junction, J. Embryol. Exp. Morphol. 35: 267 (1976).

31. French, V., Bryant, P. J. and Bryant, S. V., Pattern regulation in epimorphic fields, Science 193: 969 (1976).

32. Gehring, W., Bildung eines vollstandigen Mittelbeins mit Sternopleura in der Antennenregion bei der Mutante Nasobemia (Ns) von Drosophila melanogaster, Arch. Julius Klaus-Stift. Vereb. Forsch 41: 44 (1966)

33. Gehring, W. and Nöthiger, R., The imaginal discs of Drosophila, in Developmental Systems: Insects, S. J. Counce and C. Waddington, editors, Academic Press, New York, 212, 1973.

34. Gierer, A. and Meinhardt, H., A theory of biological pattern formation, Kybernetik 12: 30 (1972).

35. Gmitro, J. I. and Sciven, L. E., in Intracellular Transport, K. B. Warren, editor, Academic Press, New York, 1966.

36. Harrison, R. G., Experiments on the development of the forelimb of Amblystoma, a self-differentiating equipotential system, J. Exp. Zool. 25: 413 (1918).

37. Harrison, R. G., On relations of symmetry in transplanted limbs, J. Exp. Zool. 32: 1 (1921).

38. Harrison, R. G., Organization and Development of the Embryo, S. Wilens, editor, Yale University Press, New Haven, 1969.

39. Haynie, J. L. and Bryant, P. J., Intercalary regeneration in imaginal discs of Drosophila melanogaster, Nature 259: 659 (1976).

40. Haynie, J. L. and Schubiger, G., Absence of distal to proximal intercalary regeneration in imaginal wing discs of Drosophila melanogaster, Dev. Biol. 68: 151 (1979).

41. Herschkowitz-Kaufman, M., Bull. Math. Biology 37: 589 (1975).

42. Hierneux, J. and Erneux, T., Chemical patterns in circular morphogenetic fields, Bull. Math. Biol. 41: 461 (1979).

43. Holder, N., Tank, P. and Bryant, S. V., Regeneration of symmetrical forelimbs in the axolotl, Ambystoma mexicanum, Dev. Biol. 74: 302 (1980).

44. Hunt, R. K., Developmental programming for retinotectal patterns, in Cell Patterning, Ciba Foundation Symposium 29, Associated Scientific Publishers, Amsterdam, 1975.

45. Jacobson, M., Development of neuronal specificity in retinal ganglion cells of Xenopus, Dev. Biol. 17: 202 (1968).

46. Jurgens, G. and Gateff, E., Pattern specification in imaginal discs of Drosophila melanogaster: Developmental analysis of a temperature-sensitive mutant producing duplicated legs, Wilhelm Roux's Archives 186: 1 (1979).

47. Kauffman, S. A., A Cartesian coordinate model of positional information in imaginal discs of Drosophila, 20th Annual Drosophila Conference, 1978.

48. Kauffman, S. A. and Ling, E., Regeneration by complementary wing disc fragments of Drosophila melanogaster, to appear in Dev. Biol.

49. Kauffman, S. A., Shymko, R. and Trabert, K., Control of sequential compartment formation in Drosophila, Science 199: 259 (1978).

50. Kernevez, J.-P., Enzyme Mathematics, in press, Amsterdam, North Holland.

51. Krivi, G. G. and Schneiderman, H. A., Pattern regulation in Drosophila wing disc reaggregates, submitted for publication.

52. Lacalli, T. C. and Harrison, L. G., The regulatory capacity of Turing's model for morphogenesis, with application to slime moulds, J. Theor. Biol. 70: 273 (1978).

53. Meinhardt, H. A., A model of pattern formation in insect embryogenesis, J. Cell. Sci. 23: 117 (1977).

54. Milojevic, B. D., Beitrage zur Frage uber die Determination der Regenerate, Wilhelm Roux's Archives 103: 80 (1924).

55. Morgan, T. H., Regeneration, MacMillan, New York, 1901.

56. Needham, A. E., in Regeneration in Animals and Related Problems, V. Kiortsis and H. Trampusch, editors, Amsterdam, North Holland, 283, 1965.

57. Nicolis, G., Erneux, T. and Herschkowitz-Kaufman, M., Pattern formation in reacting and diffusing systems, Adv. in Chem. Phys. 38: 263 (1978).

58. Nicolis, G. and Prigogine, I., Self-organization in nonequilibrium systems, Interscience, New York, 1977.

59. Ouweneel, W. J., Developmental genetics and homeosis, Adv. Gen. 18: 179 (1976).

60. Ouweneel, W. J., Morphology and development of loboid-ophthalmoptera, a homeotic strain in Drosophila melanogaster, Wilhelm Roux's Archives 164: 1 (1969).

61. Postlethwait, J. H., Development of the temperature sensitive homeotic mutant eyeless ophthalmoptera of Drosophila melanogaster, Dev. Biol. 36: 212 (1974).

62. Postlethwait, J. H. and Schneiderman, H. A., Developmental genetics of Drosophila imaginal discs, Ann. Rev. Genetics 7: 381 (1974).

63. Reinhardt, C., Hodgkin, N. M. and Bryant, P. J., Wound healing in imaginal discs of Drosophila. I. Scanning electron microscopy of normal and healing wing discs, Dev. Biol. 60: 238 (1977).

64. Roberts, P., Mosaics involving aristapedia, a homeotic mutant of Drosophila melanogaster, Genetics 49: 593 (1964).

65. Rose, S. M., Tissue-arc control of regeneration in the amphibian limb, Symp. Soc. Study Develop. Growth 20: 153 (1962).

66. Russell, M., A spherical coordinate model of positional information, 20th Annual Drosophila Conference, 1978.

67. Sander, K., Current understanding of cytoplasmic control centers, in Insect Embryology, S. W. Visscher, editor, Montana State University Press, 31, 1977.

68. Sattinger, D. H., Topics in stability and bifurcation theory, in Lecture Notes in Mathematics, Vol. 309, Springer-Verlag, Heidelberg, 1972.

69. Schubiger, G., Regeneration, duplication and transdetermination in fragments of the leg disc of Drosophila melanogaster, Dev. Biol. 67: 386 (1971).

70. Schubiger, G. and Schubiger, M., Distal transformation in Drosophila leg imaginal disc fragments, Dev. Biol. 67: 286 (1978).

71. Sengel, P., Arch. d'Anat. Micr. 42: 57 (1953).

72. Slack, J., A serial threshold theory of regeneration, J. Theor. Biol. 82: 105 (1980).

73. Slack, J. and Savage, S., Regeneration of reduplicated limbs in contravention of the complete circle rule, Nature (London) 271: 760 (1978).

74. Slack, J. and Savage, S., Cell 14: 1 (1978).

75. Stern, C., Developmental genetics of pattern, in Genetic Mosaics and Other Essays, Harvard University Press, Cambridge, Mass., 735, 1968.

76. Stocum, D. L., Regulation after proximal or distal transposition of limb regeneration blastemas and determination of the proximal boundary of the regenerate, Dev. Biol. 45: 112 (1975).

77. Stocum, D. L., Regeneration of symmetrical hindlimbs in larval salamanders, Science 200: 790 (1978).

78. Strub, S., Development potential of the cells of the male foreleg disc of Drosophila, Wilhelm Roux's Archives 181: 309 (1977).

79. Swett, F. H., Determination of limb axes, Quart. Rev. Biol. 12: 322 (1977).

80. Swett, F. H., Exceptions to Bateson's rules of mirror symmetry, Anat. Rec. 28: 63 (1924).

81. Tank, P. W. and Holder, N., The effect of healing time on the proximodistal

organization of double-half forelimb regenerates in the axolotl, *Abystoma mexicanum*, *Dev. Biol.* 66: 72 (1978).

82. Tokunaga, C., Genetic mosaic studies of pattern formation in *Drosophila melanogaster*, with special reference to the prepattern hypothesis, in *Results and Problems of Cell Differentiation*, Vol. 9, *Genetic Mosaics and Cell Differentiation*, A. Gehring, editor, Springer-Verlag, Heidelberg, New York, 157, 1978.

83. Turing, A. M., *Philos. Trans. R. Soc. London, Ser. B.* 237:37 (1952).

84. Van der Meer, J. M. and Ouweneel, W. J., Differentiation capacities of the dorsal mesothoracic (haltere) disc of *Drosophila melanogaster*. II. Regeneration and duplication, *Wilhelm Roux's Archives* 174: 361 (1974).

85. Weiss, P., *Principles of Development*, Holt, Reinhart and Winston, New York, 1939.

86. Winfree, A. T., *The Geometry of Biological Time*, *Biomathematics*, Vol. 8, Springer-Verlag, Heidelberg, New York, 1980.

87. Wolpert, L., Positional information and pattern formation, in *Current Topics in Developmental Biology*, Vol. 6, A. A. Moscana and A. Monroy, editors, Academic Press, New York, 183, 1971.

88. Wolpert, L., Positional information and the spatial pattern of cellular differentiation, *J. Theor. Biol.* 25: 1 (1969).

BLOOD FLOW IN THE HEART

M. F. McCracken
Department of Mathematics, Indiana University,
Bloomington, Indiana

ABSTRACT

M. F. McCracken and C. S. Peskin have developed a numerical method for solving the incompressible Navier-Stokes equations at high Reynolds number in a region with moving boundary whose motion is not known in advance. The method is a hybrid vortex-grid method. The presence of the vortices means that the mesh does not have to be refined as fast as R increases as it would in a pure finite difference method. The mesh is used to keep the number of vortices from increasing as the computation proceeds. This numerical method has been applied to the problem of blood flow through the mitral valve of the heart. It has been possible to do calculations using Reynolds numbers in the range of mammalian physiology. The results of this work appear in: McCracken, M. F. and Peskin, C. S., A Vortex Method for Blood Flow Through Heart Valves, J. Comput. Physics 35: 2 (1980); McCracken, M. F., Applications of Mathematics to Problems Concerning the Performance of Heart Valves, Some Mathematical Questions in Biology 19, American Math. Society, to appear, 1980.

CONTROL AND ACTIVATION OF THE COMPLEMENT SYSTEM

Stephen J. Merrill
Department of Mathematics, Statistics and Computer Science
Marquette University, Milwaukee, Wisconsin

ABSTRACT

One aspect of the immune response to a foreign substance (antigen), is the production of proteins (antibody) specific for the antigen. If antigens lie on cell surfaces, the resulting antigen-antibody complexes may trigger the activation of a sequence of serum proteins, the complement system. This activation goes through stages of recognition, enzyme cascade and finally membrane attack with the end result of cell destruction (lysis).

The control and activation of this system is investigated via a mathematical model, especially with respect to the role of C2 concentration and the action of C1 inhibitor.

INTRODUCTION

The immune system consists of several powerful amplification mechanisms developed to recognize and eliminate possibly harmful substances. The system has two specific modes of response, the _humoral response_ and the _cell mediated response_. If any substance triggers a specific response, it is called an _antigen_. Each mode of response has a class of white cells, lymphocytes, as main participants. Each lymphocyte has receptors on its surface which can recognize (bind) molecules possessing certain shapes. The wide differences in the patterns recognized by the individual lymphocytes makes it highly unlikely that a harmful substance would fail to be recognized by some population of lymphocytes and, if in great enough concentration, initiate at least one mode of a specific response.

If a humoral response is initiated, one result is the proliferation of some of the B-lymphocytes able to bind the antigen into clones whose cells have specificity identical to that of the B-lymphocytes which recognized the antigen. The cells in the clones then either differentiate into _plasma cells_ or _memory cells_. Memory cells (which live as long as 50 years) amplify the original response if that (or a similar) antigen is again encountered. The presence of memory cells is a sign of immunity to that antigen. These memory cells apparently play no part in the ongoing humoral response. The plasma cells, however, are only living for this response (so their lifetime is only a few weeks). This cell can produce enormous quantities of a protein, _antibody_, which has the ability to bind to the same molecular pattern as the original B-lymphocyte which generated its clone.

The antibody produced by the humoral response is dumped into the blood and lymph where it binds to the antigen and thus tags the antigen as something that should be removed. For antigens carried on viruses or toxins, this binding may in itself be sufficient to remove the harmful nature of the virus or toxin. For other antigens especially those on the surface of bacteria, the binding often fails to remove the harmful potential.

Mammals have two mechanisms to utilize this antibody on a cell surface to effect the removal of the bacteria. One is the complement system and the other is opsonization. Activation of the complement system can result in cell destruction (lysis) while opsonization results in a more specific and rapid removal of the tagged material by phagocytic cells, macrophages and granulocytes [9].

The complement system as currently understood can utilize one of two principal activation sequences, the classical pathway and the alternative pathway. Each of these pathways consists of four serum proteins (one of which is used for control of the activation). The final serum proteins (one of which is a control protein) are utilized if either one of the activation sequences has been completed. There are many other ways *in vitro* to avoid various steps in the system, but they have not been shown to be operable *in vivo* [8]. The classical pathway for activation is initiated by the first complement component, C1, binding to certain regions of two class IgG antibody molecules which are bound to an antigen on a cell surface or one IgM class antibody molecule bound to a cell surface antigen. The result of the C1 binding is an enzyme, $C1\bar{s}$, which has both proteins C4 and C2 as substrates. A control protein, the C1-inhibitor or C1-INH, regulates $C1\bar{s}$. The splitting of C4 and C2 and the resulting binding of active fragments marks the end of the classical pathway of activation. The alternative pathway is less well understood and seems to be activated directly by certain antigens.

Following activation by either the classical or alternative pathway (which is not antibody dependent), the result is a C3-convertase, an enzyme which has C3 as a substrate. The amplification involved in the production of this enzyme along with the biologic consequences of the splitting of the C3 component make the control of the initial activation and amplification of great importance.

Lack of the usual control mechanisms is a causal factor in several diseases (e.g., HANE), and perturbations of these mechanisms as seen in many other diseases (e.g., systemic lupus erythematosis) is responsible for much of the clinical problems experienced by those individuals.

In this paper a mathematical model of the complement system developed in [11] and [12] will be used to explore the mechanisms present in the classical pathway which control production of the C3-convertase, $\overline{C4b2a}$. Special attention is given to the C1 inhibitor, $C\overline{1}$-INH, and the local control exercised by the production of C2. An excellent textbook of immunology is [3], and a detailed survey of the various aspects of the complement system (as well as opsonization) can be found in [3] and [5]. An outline of the humoral response and several mathematical models proposed is [10], and an overview of theoretical immunology can be found in [2].

<u>Section 1</u>. The classical pathway of complement activation.

The activation of the complement system by the classical pathway depends on the presence of antigen-antibody complexes. The usual sequence of events is

1) cell-bound antigens trigger a humoral immune response, resulting in the production of antibodies, specific for the antigens.

2) the antibody produced during a humoral response (initially class IgM and later class IgG) binds to their target antigens resulting in antigen-antibody complexes being present on cell surfaces.

3) the first protein of the complement system, C1, attaches to a region in the "Fc" portion of the antibody on the complex. The activation of C1 requires that two of the six heads of the C1q portion of the molecule bind to Fc sites on antibody. This requires two IgG class antibodies in close proximity or one IgM class antibody. The structure of IgG and IgM are shown in Figures 10.1 and 10.2. The structure of C1 molecule as it is activated by IgG molecules is shown in Figure 10.3.

As a result of the binding of the C1q unit of the C1 molecule, the C1r unit of C1 changes its conformation, eventually presenting $C1\overline{r}$ (the activated form of C1r). $C1\overline{r}$ is an enzyme with substrate the C1s portion of the C1 molecule. The cleavage of C1s by $C1\overline{r}$ produces $C1\overline{s}$, an enzyme with both C4 and C2 as substrates.

The splitting of C4 into C4a and $C4\overline{b}$, the active form, results in roughly 10% of the $C4\overline{b}$ binding to a cell surface. The cell binding site on $C4\overline{b}$ is very short lived and thus if $C4\overline{b}$ does bind to cell membrane, it must be very close to the site of C4 splitting. The membrane bound $C4\overline{b}$ provides a site for the binding of $C2\overline{a}$ (one of the products of $\overline{C1s}$ splitting C2). $C4\overline{b}$ present near $C1\overline{s}$, either on a membrane or possibly Cb4 with an inactive membrane binding site still bound to $C1\overline{s}$, enhances the splitting of C2 by $C1\overline{s}$. Proposed mechanisms include $C4\overline{b}$ loosely binding C2 so that $C1\overline{s}$ can more easily split C2 (by increasing substrate concentration near the enzyme) or by $C4\overline{b}$ exposing a $C1\overline{s}$ site on C2 or by $C4\overline{b}$ changing $C1\overline{s}$ specificity through allosteric modification of $C1\overline{s}$. A final proposal is that C4 binds

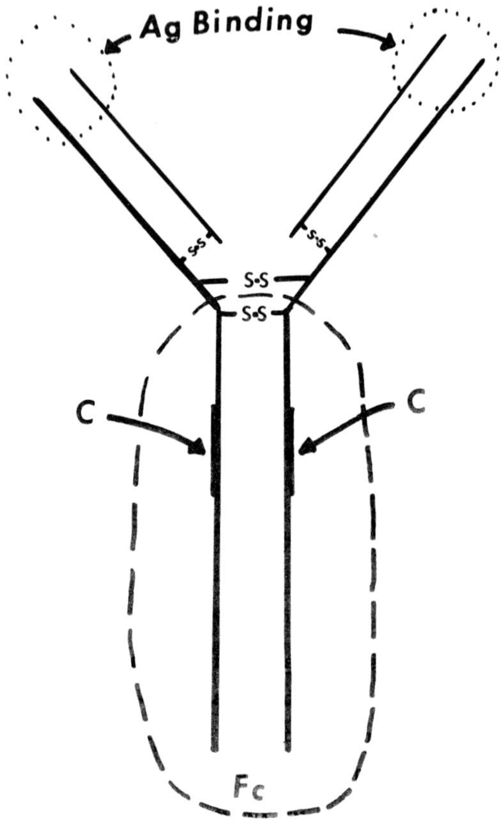

Fig. 10.1. Diagram of the structure of one IgG class antibody molecule. The C's mark the sites of complement binding and s-s's give the position of some of the disulfide bonds which hold the chains together.

the $C2\bar{a}$ product, removing it from vicinity of $C1\bar{s}$ so that $C1\bar{s}$ can interact with additional C2. As this mechanism is not fully elucidated, we will assume the reaction

$$C1\bar{s} + C2 \rightarrow C2\bar{a} + C2b + C1\bar{s}$$

has a rate which is a nondecreasing function of C4.

The active product of C2 splitting by $C1\bar{s}$ is $C2\bar{a}$ which binds to membrane bound C4b (C4b mem). The $\overline{C4b2a}$ complex on a cell surface is the C3 convertase--an enzyme whose substrate is C3. The biological effects of the splitting of C3 into C3a + C3b are widespread and because of that, production of the products of splitting and the location of their production must be carefully controlled.

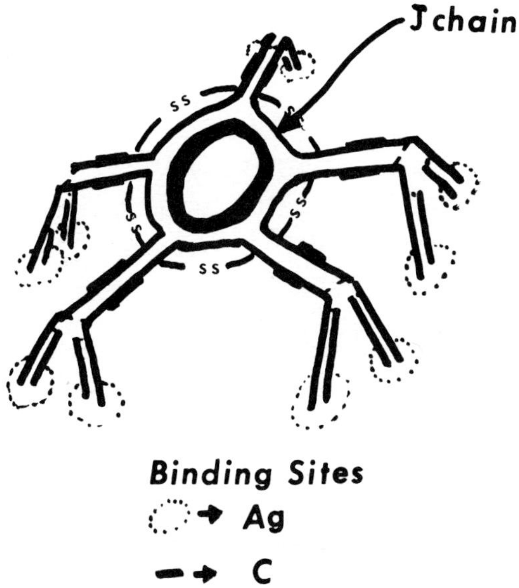

Fig. 10.2. Diagram of the structure of one IgM class antibody molecule. Its structure is not unlike five IgG antibody molecules bound by a "J" chain. The heavy line segments mark the 10 sites of possible complement binding. The sites of antigen (Ag) binding are the circled regions.

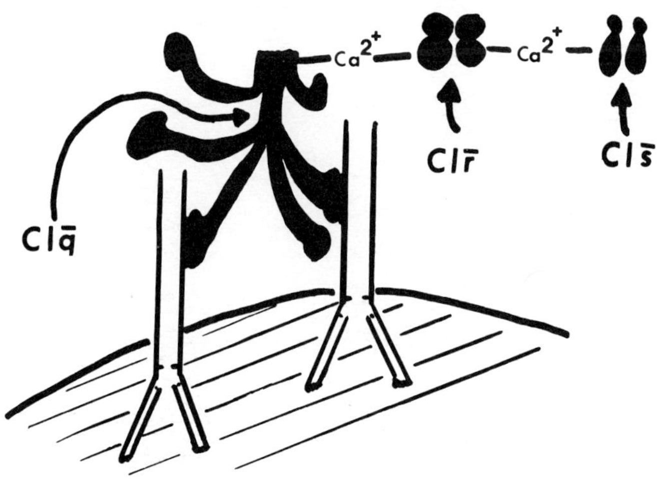

Fig. 10.3. Diagram of C1 binding to two IgG molecules. The components of C1; C1q, C1r and C1s are indicated.

A fraction of the C3b produced binds to the $\overline{C4b2a}$ complex producing $\overline{C4b2a3b}$, the C5 convertase. The $\overline{C4b2a3b}$ is very stable and shows little dissociation. The splitting of C5 by its convertase is the last enzyme mediated reaction. The $\overline{C5b}$ produced has a short-lived membrane binding site and its rapid attachment to a cell surface, either before or after C6 and C7 have been added to the complex, is necessary if the final two complement proteins, C8 and C9, are to be utilized. It is as a result of the rapid addition of C8 and up to six C9 molecules to the membrane bound $\overline{C5b67}$ complex that results in cell lysis.

Section 2. Control of the complement system.

Control of the complement system depends on an interacting set of mechanisms:

1) the quantity of components of the system and their distribution,
2) inhibitors and inactivators of enzymes produced, and
3) the rates of decay and interaction being dependent on local conditions.

1) Quantity of components and their distribution.

Table 1 contains information on the relative abundance of the initial complement components and their mode of production.

TABLE 1 ([3], [4])

Protein	Site of Production	Serum Concentration (μg/ml)	Relative Molecule Concentration (C1q = 1)
C1q	intenstine	190	1
C2	macrophage	20-30	.3
C4	macrophage & liver	400-430	4
C3	liver	1200-1300	12
C1-INH	liver	105-180	4
C3b-INA	?	40-50	1

It is important to note from Table 1 that the only site of C2 production so far discovered is the macrophage, a phagocytic white cell widely distributed in tissue and blood. All other components are generally stable and produced by organs which dump their product in the blood, and thus the distributions of all components, except C2 (and possibly C4) are roughly constant spacially. C2, however, is also unstable *in vivo*, and, as a result, active C2 is available in higher concentrations near sources. An additional piece of evidence for spacially heterogeneous C2 (and possibly C4) concentrations is provided by studies which show that normal guinea pig macrophages when exposed to pneumococci *in vitro* stimulated C2 and C4 biosynthesis ten fold [4]. There are also serum factors which have a stimulation ability which has not yet been clearly defined. The crucial fact is, however, that macrophages are able to adjust the synthesis rate of C2 and C4 to respond to local conditions.

2) *Inhibitors and inactivators.*

The inhibitors and inactivators responsible for control of the complement system are:

C1-INH (C1 inhibitor). C1-INH irreversibly binds to $C1\overline{r}$ and $C1\overline{s}$. C1-INH is also an inhibitor for enzymes produced in other plasma protein systems.

C3-INA (C3 inactivator). C3-INA is an enzyme that splits and thus inactivates $C3\overline{b}$, $C4\overline{b}$ and $\overline{C5b6}$. As C3-INA is an enzyme and is not expended during inactivations, C3-INA serves the important function of removing $C3\overline{b}$, $C4\overline{b}$ and $\overline{C5b6}$ once the reaction which produces these products is over, sending concentrations of these active components to zero.

AI (anaphylatoxin inactivator). Eliminates the biologic activity of C3a and C5a, two fragments produced which do not further influence the complement system. AI, as C3-INA above, is an enzyme. Its action, once production of C3a and C5a has ceased, sends the concentration of these fragments to zero.

S protein. Binds to $\overline{C5b67}$ to prevent cellular attachment. The complex continues to add C8 and C9 although no cytolytic reaction results.

Beta - 1-H. Functions to augment the rate of inactivation of C3b by C3b-INA. It also acts in the alternative pathway which will not be discussed here.

The actions of Beta - 1-H and C3-INA are not sufficient to prevent the complement system from progressing. Their role is to modify the production of $\overline{C3b}$ after activation and hasten $\overline{C3b}$ inactivation until it is no longer present. If the C3 inactivators were not present, each activation would continue until all C5 was expended.

In [11] it was shown that the C1-INH acts to provide a threshold in activation. Unless there was sufficient antigen antibody complexes present, no amplification would result. This is demonstrated in Section 4.

3) Rates of decay and interaction.

We have already discussed the result of the production and stability of C2 in providing spacial difference in C2 concentrations. In addition, the membrane binding sites of $\overline{C4b}$ and $\overline{C5b}$ are both short lived and the probability of their binding to a membrane is a function of membrane availability (cell density). The C3 convertase, $\overline{C4b2a}$, also has a half-life of about 10 minutes at 37° C. The C2a portion inactivates and enters the fluid phase, and the $\overline{C4b}$ remains at the membrane able to again bind C2a.

This mechanism guarantees that C3 will be split only where C2 is available. This is available near sources (the macrophages) which are most active when stimulated to increase C2 and C4 production by serum factors or by macrophage contact with the antigen. When the antigen has been removed and the response should end, $\overline{C4b}$ on the membrane is unable to obtain C2a which results in no more C3 being split. Now the inactivators take over, and the response is quickly terminated with control that is all localized.

Once the response has gone past $\overline{C5b67}$ on a membrane, no control mechanism exists, and it will terminate in lysis unless C8 or C9 is depleted.

Section 3. A model of the classical pathway of complement activation.

The model is a system of differential equations based on the analysis of the kinetics of enzymes present and other mechanisms unique to the system. Similar models and background are available in [1]. Details for the construction of the equations can be found in [12] and a mathematical analysis of the first part of the model appear in [11].

Preliminaries for the model

notation:

$\frac{dF}{dt}$ = rate of change in time of the quantity F

[z] = concentration of z

A∿B = A-B complex

$\overline{T}$ = activated form of T

CAb_G = concentration of complement fixing antigen-IgG complex (Ag∿IgM)

CAb_M = concentration of complement fixing antigen-IgM complex (Ag∿IgM)

B = concentration of $\overline{C1}$-inhibitor ($\overline{C1}$-INH)

$B_1 = [C\bar{1}\text{-INH}\sim C1\bar{r}]$

$B_2 = [C\bar{1}\text{-INH}\sim C\bar{1}s]$

C = concentration of $C3\bar{b}$-inactivator ($C3\bar{b}$-INA)

For J = 1, 2, 3, 4, 5, 6, 7, 8, 9

CJ = concentration of the complement component numbered J

CJu = concentration of the component numbered J used and unavailable for further reactions

All constants are assumed to be greater than or equal to zero, and initial values of all time changing values are assumed greater than or equal to zero.

A model of the recognition unit transitions and reactions and their rates

1a. $2(\text{Ag}\sim\text{IgG}) + C1q \underset{k_{-1}}{\overset{k_1}{\rightleftarrows}} 2(\text{Ag}\sim\text{IgG})\sim C1\bar{q}_G$

1b. $(\text{Ag}\sim\text{IgM}) + C1q \underset{k^*_{-1}}{\overset{k^*_1}{\rightleftarrows}} (\text{Ag}\sim\text{IgM})\sim C1\bar{q}_M$

2. $C1r + C1\bar{q} \xrightarrow{k_2} C1\bar{r} + C1\bar{q}$ (Ca^{2+} flow)

We assume that $C1\bar{q}$ only activates the C1r on the same molecule.

3. $C1\bar{r} + C1\bar{q} \xrightarrow{k_{-2}} C1r + C1q$

4. $C1\bar{r} + C\bar{1}\text{-INH} \xrightarrow{k_3} C1\bar{r}\sim C\bar{1}\text{-INH}$

Reaction 4 is not reversible

5. $C1s + C1\bar{r} \xrightarrow{k_4} C1\bar{s} + C1\bar{r}$

We assume that $C1\bar{r}$ only activates C1s on the same molecule.

6. $C1s \xleftarrow{k_{-4}} C1\bar{s}$

7. $C1\bar{s} + C\bar{1}\text{-INH} \xrightarrow{k_5} C1\bar{s}\sim C\bar{1}\text{-INH}$

Reaction 7 is not reversible.

Equations:

$$(10.1)\quad \frac{d[C1\bar{q}_M]}{dt} = k_1^*(C1 - ([C1\bar{q}_M] + [C1\bar{q}_G]))\,(CAb_M - [C1q_M]) - k^*_{-1}\,[C1\bar{q}_M]$$

$$(10.2)\quad \frac{d[C1\bar{q}_G]}{dt} = k_1^*(C1 - ([C1\bar{q}_M] + [C1\bar{q}_G]))\,(CAb_G - 2[C1\bar{q}_G])^2 - k_{-1}[C1\bar{q}_G]$$

(10.3) $\overline{C1q} = [\overline{C1q}_M] + [\overline{C1q}_G]$

(10.4) $$\frac{d[\overline{C1r}]}{dt} = k_2(\overline{C1q} - B_1 - [\overline{C1r}]) - k_{-1}\frac{[\overline{C1q}_G]}{\overline{C1q}}[\overline{C1r}] - k^*_{-1}\frac{[\overline{C1q}_M]}{\overline{C1q}}[\overline{C1r}] - K_3[\overline{C1r}](B - (B_1 + B_2))$$

(10.5) $$\frac{d[\overline{C1s}]}{dt} = k_4([\overline{C1r}] - B_2 - [\overline{C1s}]) - k_{-4}[\overline{C1s}] - k_5[\overline{C1s}]\ (B - (B_1 + B_2))$$

(10.6) $$\frac{dB_1}{dt} = k_3[\overline{C1r}](B - (B_1 + B_2))$$

(10.7) $$\frac{dB_2}{dt} = k_5[\overline{C1s}](B - (B_1 + B_2))$$

Discussion:

(10.1) & (10.2) $\overline{C1q}$ production depends on C1 that has not been activated, $(C1 - ([\overline{C1q}_M] + [\overline{C1q}_G]))$, and the availability of antigen-antibody complexes, either of type IgM or IgG. Each $\overline{C1q}$ activated by IgM, $\overline{C1q}_M$, occupies one Ag-IgM unit. Each $\overline{C1q}$ activated by IgG, $\overline{C1q}_G$, occupies two Ag-IgG units. Both $\overline{C1q}_M$ and $\overline{C1q}_G$ can revert to C1q by dissociation with the antigen-antibody complex.

(10.3) The concentration of activated C1q $= [\overline{C1q}_M] + [\overline{C1q}_G]$.

(10.4) C1r is activated when C1q is activated. $\overline{C1r}$ can become C1r when $\overline{C1q}$ dissociates from antibody or when inhibited by irreversible complexing with $\overline{C1}$-INH.

(10.5) C1s can be activated when the C1r on the same molecule has become activated. $\overline{C1s}$ can be inhibited by irreversible binding to $\overline{C1}$-INH or can revert to C1s when $\overline{C1q}$ dissociates from antibody.

(10.6) & (10.7) Concentration of $\overline{C1}$-INH complexed with $\overline{C1r}$ and $\overline{C1s}$ respectively.

A model of the amplification unit

Reactions and transitions:

8. $C4 + \overline{C1s} \xrightarrow{k_6} \overline{C4b}_{f1} + C4a + \overline{C1s}$

9. $\overline{C4b}_{f1} \xrightarrow{k_{-6}} C4b$

This decay occurs rapidly if $\overline{C4b}$ does not attach to the cell membrane. We assume no decay if $\overline{C4b}$ is membrane attached.

10. $\overline{C4b}_{fl} \xrightarrow{k_7'} \overline{C4b}_{mem}$

11. $\overline{C4b}_{fl\ or\ mem} + \overline{C3b}\text{-INA} \xrightarrow{k_7} C4b_{inac} + \overline{C3b}\text{-INA}$

Actual product is C4c + C4d

12. $C2 + C1\bar{s} \xrightarrow{k_8} C2\bar{a} + C2\bar{b} + C1\bar{s}$

k_8 is a nondecreasing function of $(C4b_{fl} + C4b_{mem})$.

13. $\overline{C2a} \xrightarrow{k_{-8}} C2a_{inac}$

This decay occurs rapidly unless $\overline{C2a}$ complexes with $\overline{C4b}$.

14. $\overline{C2a} + \overline{C4b}_{mem} \xrightarrow{k_9} \overline{C4b2a}$

$\overline{C4b2a}$ is C3-convertase

15. $\overline{C4b2a} \xrightarrow{k_{-9}} \overline{C4b}_{mem} + C2a_{inac}$

The $\overline{C4b}$ remains in activated form.

16. $C3 + \overline{C4b2a} \xrightarrow{k_{10}} C3a + \overline{C3b} + \overline{C4ba}$

17. $\overline{C3b} + \overline{C4b2a} \underset{k_{-10}^{\#}}{\overset{k_{10}^{\#}}{\rightleftarrows}} \overline{C4b2a3b}$

$\overline{C4b2a3b}$ is C5-convertase. $k_{-10}^{\#}$ is very small. In the analysis of [12], $k_{-10}^{\#}$ is assumed to be zero.

18. $\overline{C3b} + \overline{C3b}\text{-INA} \xrightarrow{k_{10}^{*}} C3b_{inac} + \overline{C3b}\text{-INA}$

19. $\overline{C3b} \xrightarrow{k_{-10}} C3b$

Fraction decaying due to failure to attach to $\overline{C4b2a}$.

Equations:

(10.8) $$\frac{d[\overline{C4b}]_{fl}}{dt} = k_6(C4 - [C4u])[C1\bar{s}] - k_{-6}[\overline{C4b}]_{fl} - k_7[\overline{C4b}]_{fl}C - k_7'[\overline{C4b}]_{fl}$$

(10.9) $$\frac{d[C4u]}{dt} = k_6(C4 - [C4u])[C1\bar{s}]$$

(10.10) $$\frac{d[C4b]_{mem}}{dt} = k_7'[\overline{C4b}]_{fl} - k_9[C2a][C4b]_{mem} + k_{-9}[\overline{C4b2a}] - k_7[\overline{C4b}]_{mem}C$$

(10.11) $$\frac{d[\overline{C2a}]}{dt} = k_8(C2 - [C2u])[\overline{C1s}] - k_{-8}[\overline{C2a}] - k_9[\overline{C2a}][\overline{C4b}]_{mem}$$

(10.12) $$\frac{d[C2u]}{dt} = k_8(C2 - [C2u])[\overline{C1s}]$$

(10.13) $$\frac{d[\overline{C4b2a}]}{dt} = k_9[\overline{C4b}]_{mem}[\overline{C2a}] - k_9[\overline{C4b2a}] - k^{\#}_{10}[\overline{C3b}][\overline{C4b2a}] + k^{\#}_{-10}[\overline{C4b2a3b}]$$

(10.14) $$\frac{d[\overline{C3b}]}{dt} = k_{10}[\overline{C4b2a}](C3u]) - k_{-10}[\overline{C3b}] - k^{*}_{10}[\overline{C3b}]C$$

(10.15) $$\frac{d[C3u]}{dt} = k_{10}[\overline{C4b2a}](C3 - [C3u] - k^{\#}_{-10}[\overline{C4b2a3b}])$$

(10.16) $$\frac{d[\overline{C4b2a3b}]}{dt} = k^{\#}_{10}[\overline{C3b}][\overline{C4b2a}] - k^{\#}_{-10}[\overline{C4b2a3b}]$$

Discussion:

(10.8) & (10.9) C4 is used (C4u) when it is cleaved by $\overline{C1s}$. This leaves $\overline{C4b}_{fluid}$ to decay (rate k_{-6}), be inactivated by $\overline{C3b}$-INA or attach to a membrane at rate k'_7, which is cell density dependent.

(10.10) The membrane-bound $\overline{C4b}$ can complex with $\overline{C2a}$ or be inactivated by $\overline{C3b}$-INA.

(10.11) & (10.12) $\overline{C2a}$ is produced when $\overline{C1s}$ cleaves unused C2. $\overline{C2a}$ can complex with $\overline{C4b}_{mem}$ or decay (at rate k_{-8}).

(10.13) The $\overline{C4b2a}$ complex formed by the complexing of $\overline{C4b}$ and $\overline{C2a}$ can itself complex with $\overline{C3b}$, a reversible reaction.

(10.14) & (10.15) $\overline{C3b}$ is produced, and C3 used, by C3 contact with $\overline{C4b2a}$ (C3-convertase). $\overline{C3b}$ may decay, be inactivated by C3b-INA or complex (reversibility) with $\overline{C4b2a}$.

(10.16) $\overline{C4b2a3b}$, the C5-convertase, is produced when C3-convertase complexes with $\overline{C3b}$.

Section 4. Simulations and discussion.

As shown in the analysis of the model in [12], the behavior depends very little on the parameters, as long as they are chosen to be positive. As very few of the parameters are known, the main goal of the simulations was to illustrate the effect of several of the most important control mechanisms and not fit any particular in vivo data. Table 2 gives parameter values used in all simulations, each of which was accomplished by solving the system of differential equations repeatedly.

The purpose of the C1 inhibitor is clearly demonstrated in the disease Hereditary Angioneurotic Edema (HANE) in which individuals have an inherited deficiency of C1-INH [6]. The disease is characterized by episodic swelling of the extremities, face, abdominal viccera or airway. The therapy is based on avoidance of trauma and plasma transfusions to temporarily provide C1-INH. The edema during an attack is caused by a product of the complement components interacting with plasmin, a plasma protein also inhibited by C1-INH.

$$(C1\overline{s} + C2) + \text{Plasmin}$$

results in an increase in vascular permeability [7]. The model was used to show the effect of altering C1-INH concentration which is displayed in Figure 10.4. As was shown in [11], C1-INH concentration sets the threshold of antigen-antibody complex concentration necessary to activate the complement system. With no active C1-INH, there is no such threshold.

With C2 concentration fixed, Figure 10.4 further illustrates the effect of C1-INH concentrations for various concentrations of antigen-antibody complexes on $C1\overline{s}$ activation.

Figure 10.5 displays the effect of C2 concentrations on $C3\overline{b}$ concentration. It is easy to see that its main effect is to limit production. Figure 10.6 further illustrates this effect. Local variation in C2 concentration would thus dictate the level of C3 activation in that locality.

As was shown in [12], the effect of the C3 inhibitor is to terminate the system's response by eliminating $\overline{C3b}$. As can be seen in Figure 10.6, in all instances

$$\lim_{t\to\infty} [C3\overline{b}](t) = 0.$$

Table 2

C1 = 150	
C2 = 1600	$k_{-8} = 1.$
C4 = 640	$k_9 = 1.$
$k_1 = 10.$	$k_{-9} = 1.$
$k_{-1} = 1.$	$k_{10} = 10.$
$k_2 = 1.0E3$	$k_{-10} = 1.$
$k_3 = 10.$	$k^*_{10} = 10.$
$k_4 = 10.$	$k^{\#}_{10} = 10.$
$k_{-4} = 1.$	$k^{\#}_{-10} = 0.$
$k_5 = 10.$	
$k_6 = 10.$	
$k_{-6} = 100.$	
$k_7 = 10.$	

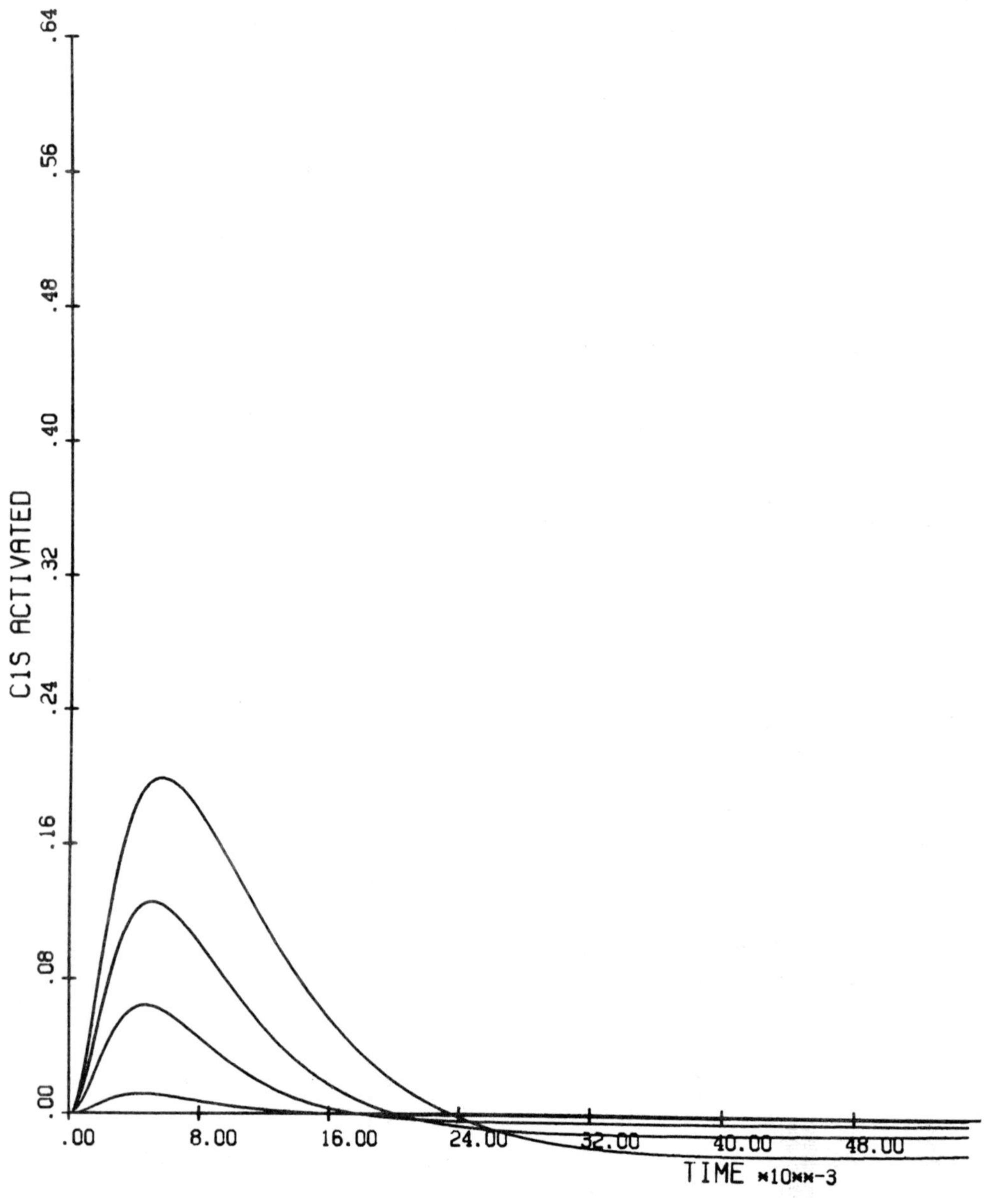

Fig. 10.4. Simulations of the model illustrating the effect of concentration of the C1 inhibitor, B. C2 concentration is held constant at 20 units. For each illustration, B is fixed and antigen-antibody complex concentration is varied. The complexes are removed at a rate proportional (with constant 1.) to $\overline{C3b}$ and complexes present. Half of complexes are IgM and half are IgG.
Fig. 10.4a. B = 40 units, total complex: 2, 10, 18, 26 units. $C1\overline{s}$ plotted.

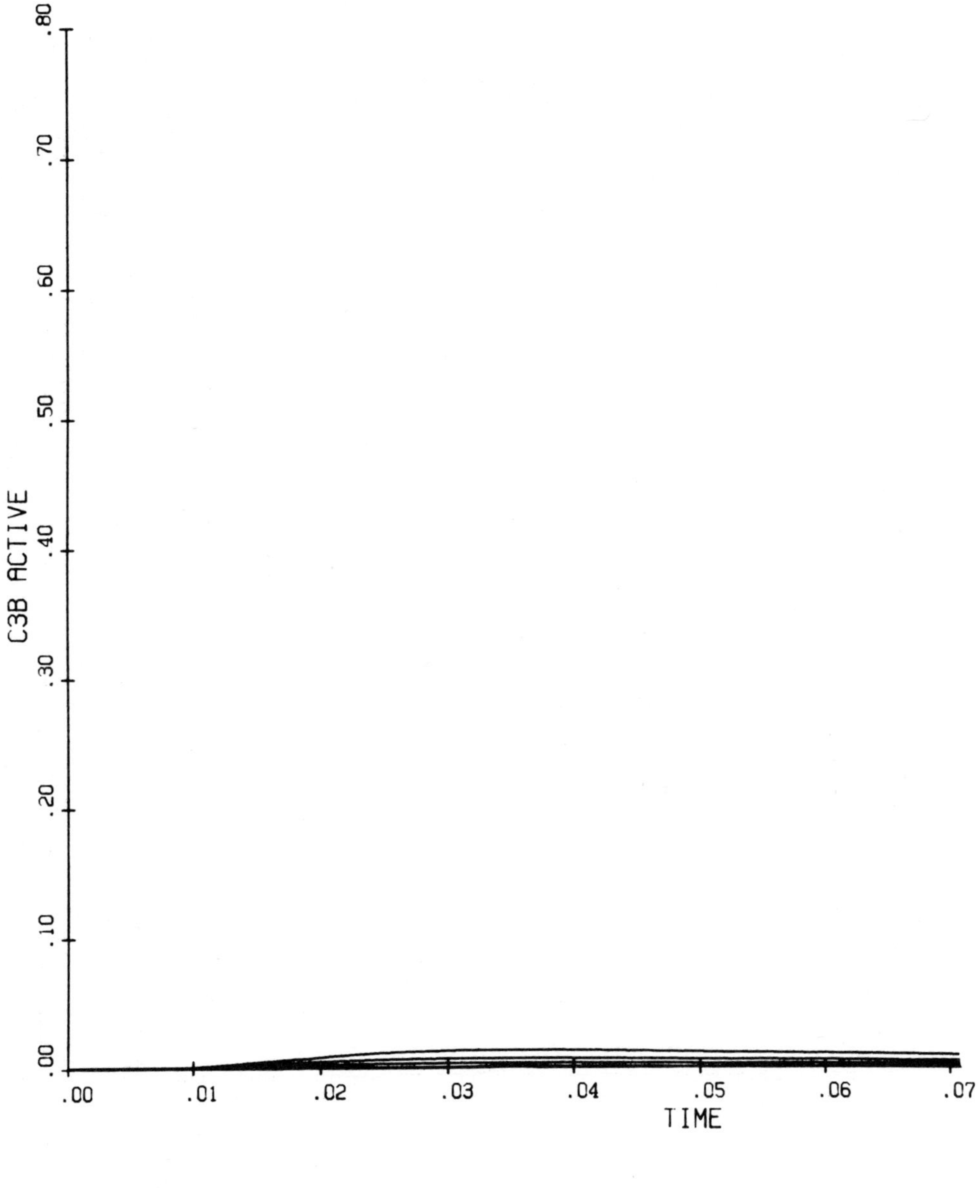

Fig. 10.4b. B = 40 units, total complex: 4, 8, 12, 16, 20. $\overline{C3b}$ plotted. (Only the highest three complex concentrations give measurable $\overline{C3b}$.

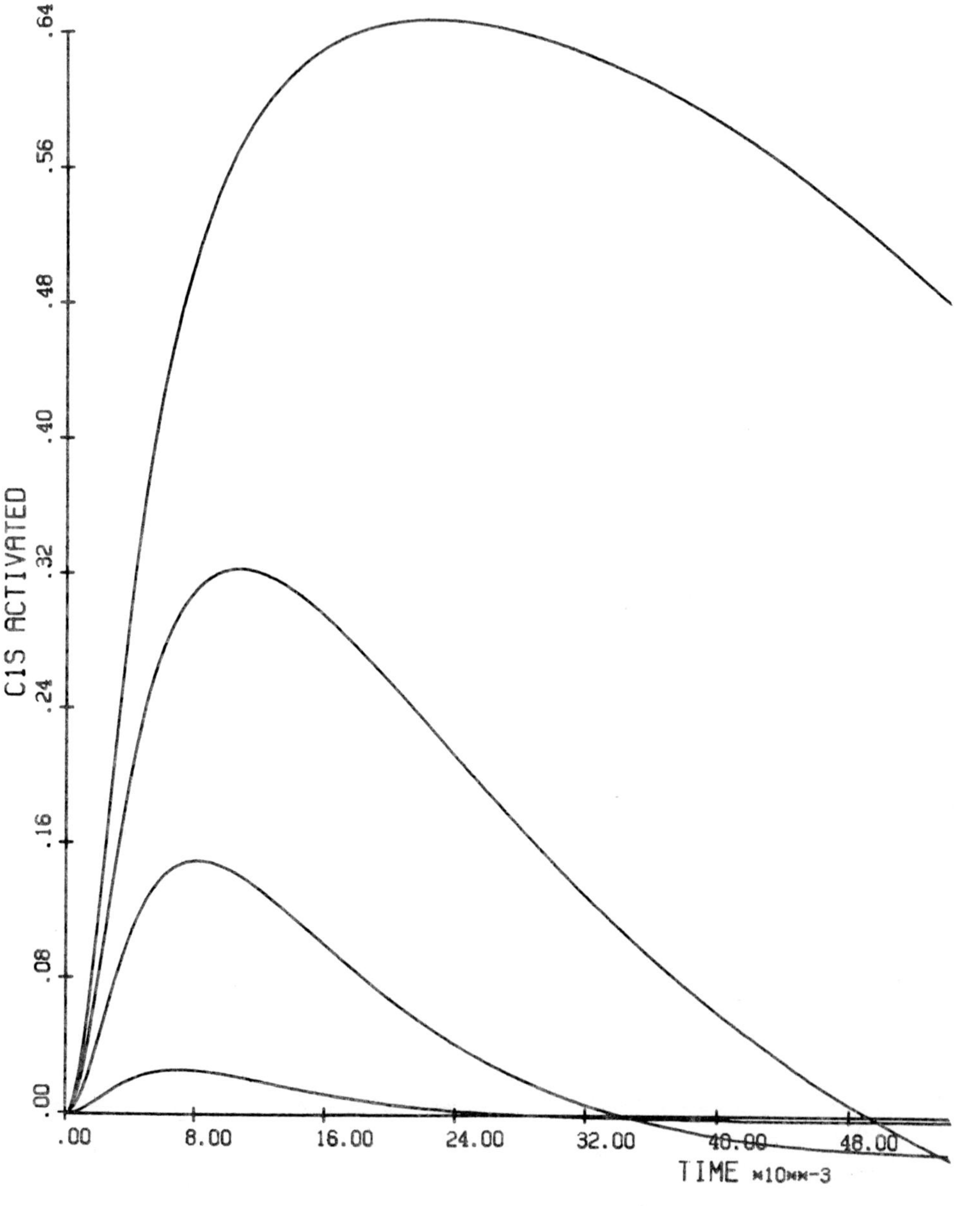

Fig. 10.4c. B = 20 units, total complex: 2, 10, 18, 26 units. $C1\bar{s}$ plotted.

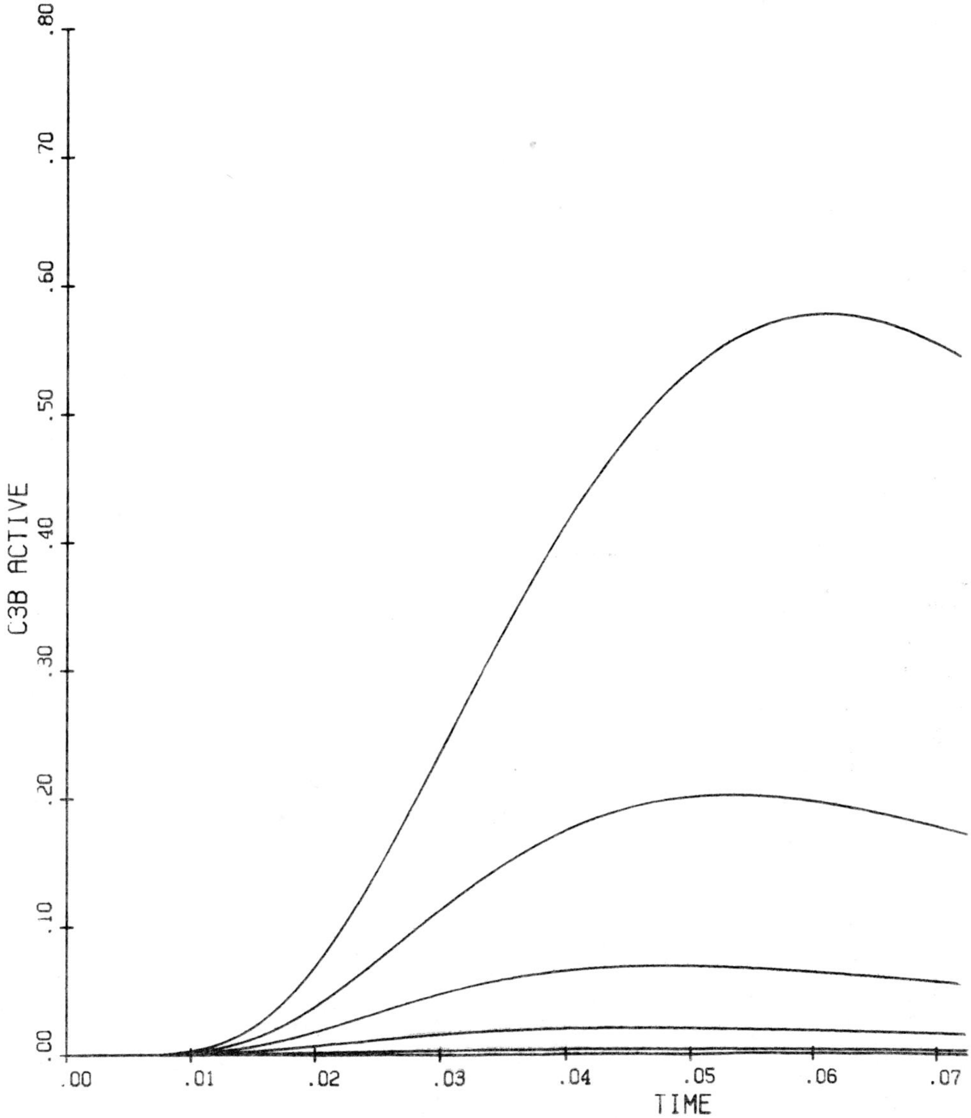

Fig. 10.4d. B = 20 units, total complex: 4, 8, 12, 16, 20 units. $C3\overline{b}$ plotted

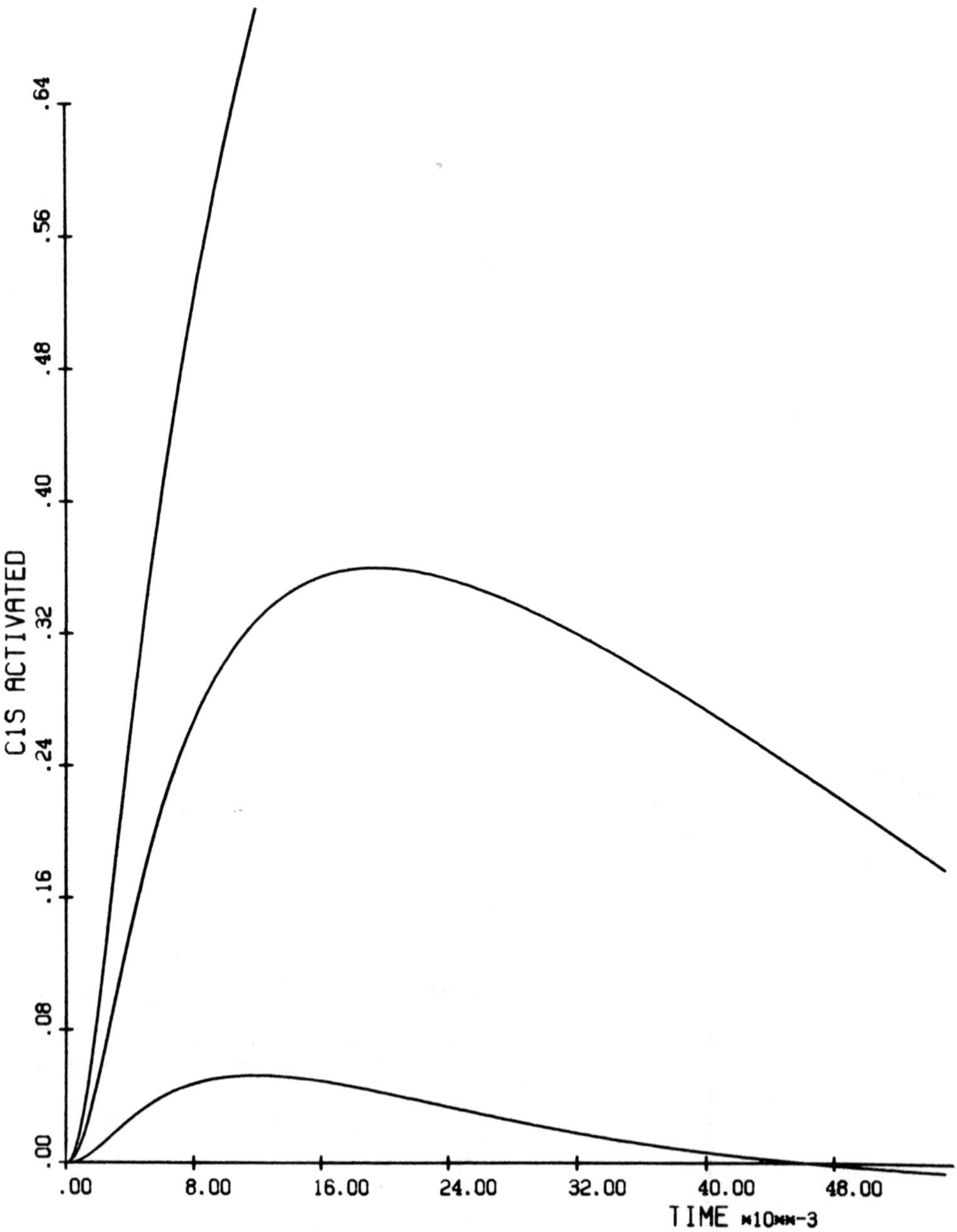

Fig. 10.4e. B = 10 units, total complex: 2, 10, 18 units. $C\overline{1s}$ plotted.

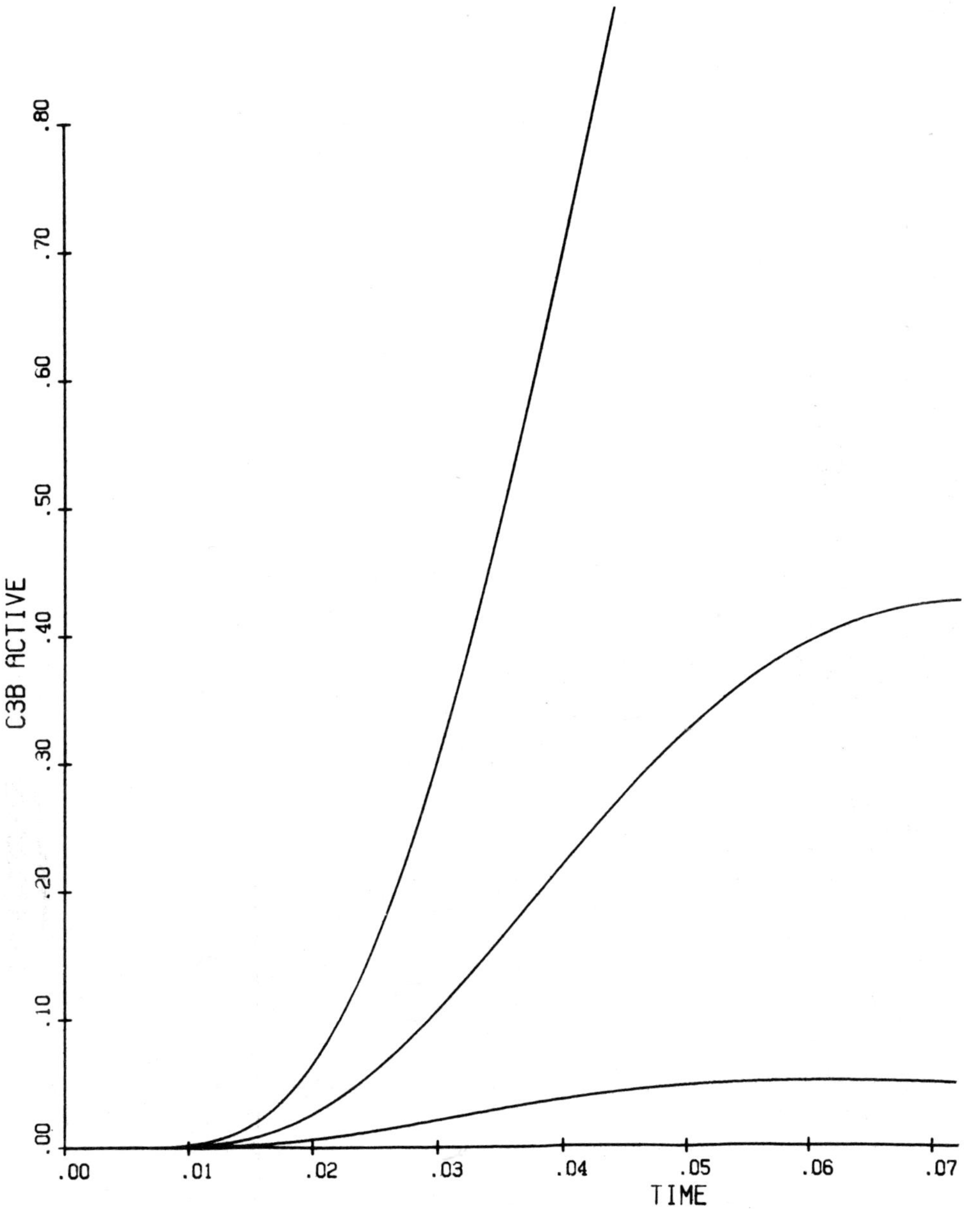

Fig. 10.4f. B = 10 units, total complex: 4, 8, 12 units. $C3\overline{b}$ plotted.

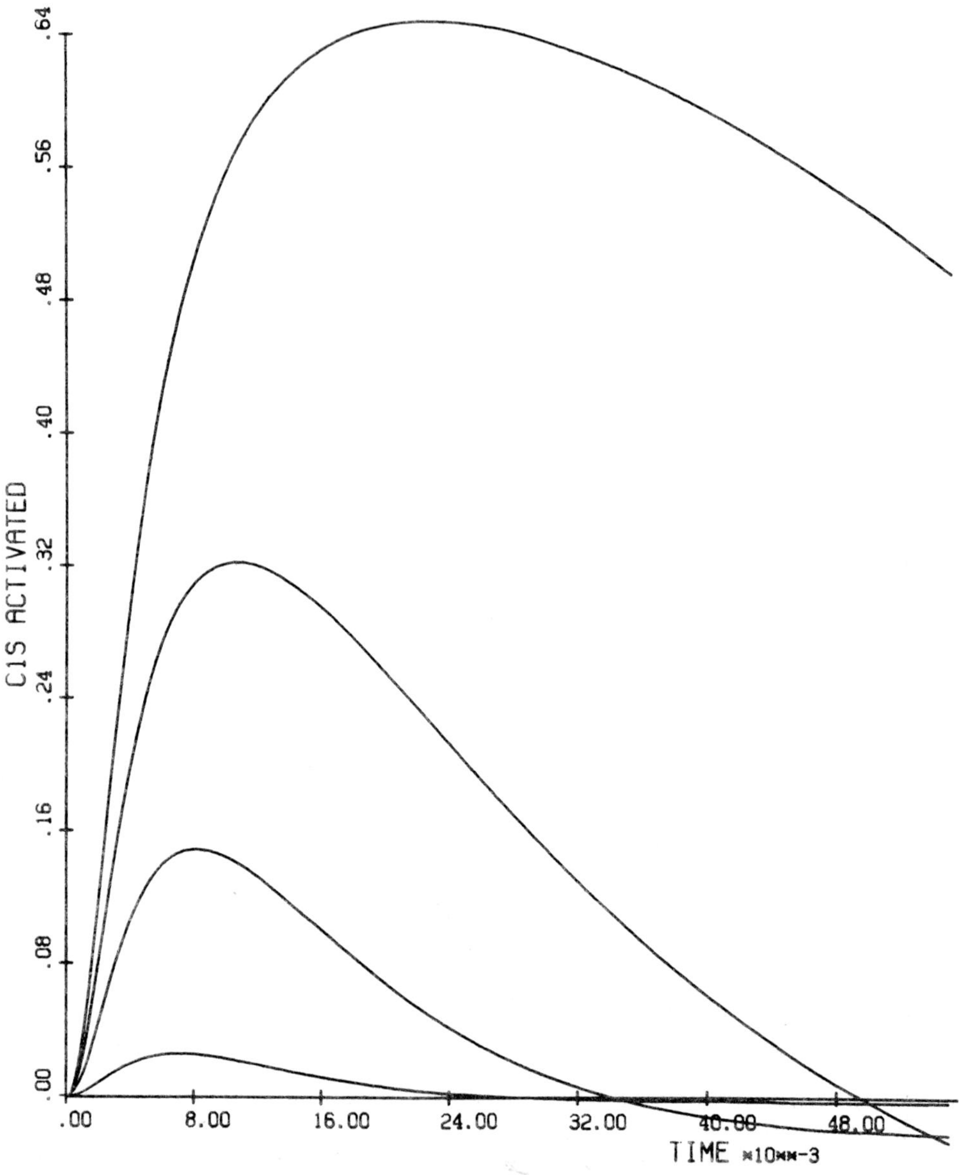

Fig. 10.5. Simulations illustrating the effect of concentration of C2. Concentration of the C1 inhibitor, B, are as noted.
Fig. 10.5a. C2 = 10, B = 20, total complex: 2, 10, 18, 26 (compare with 4c). $C\overline{1s}$ plotted.

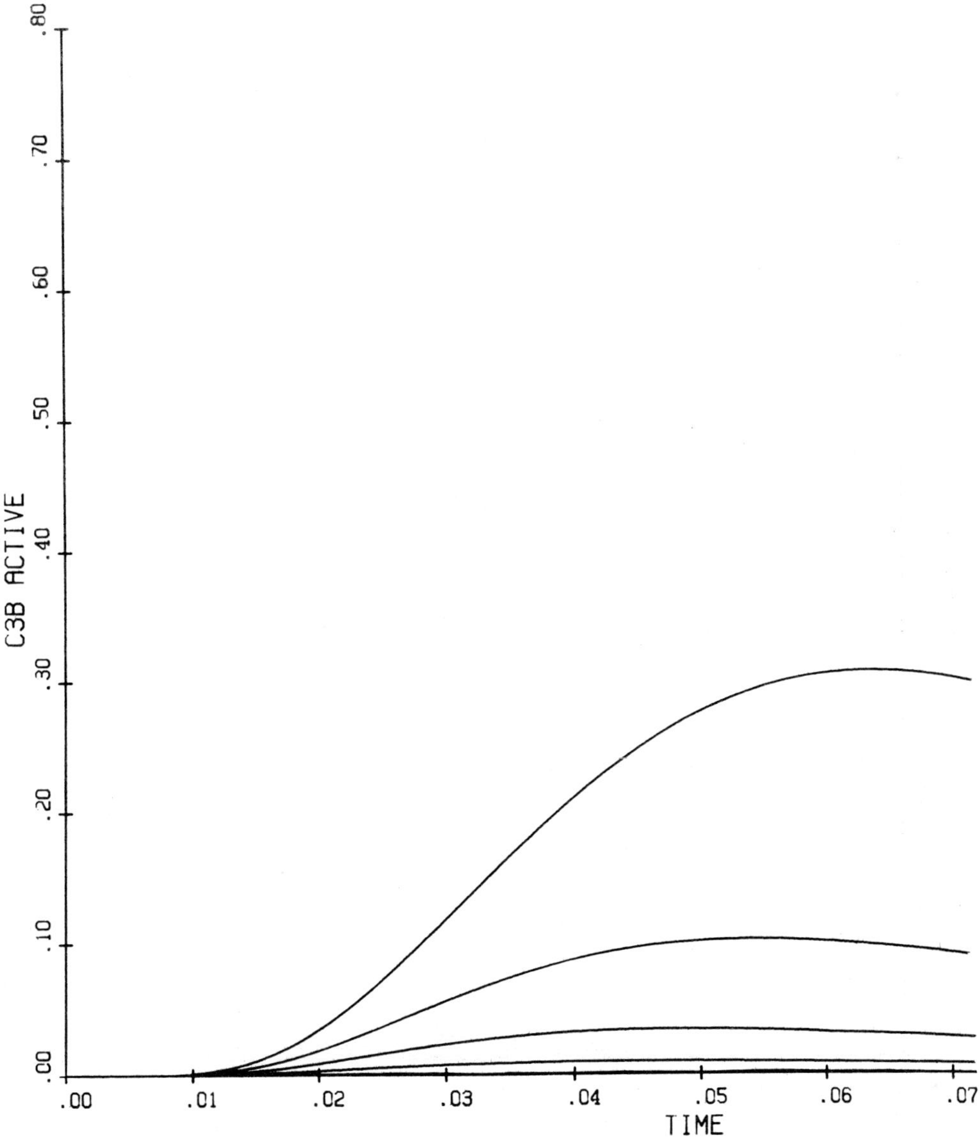

Fig. 10.5b. C2 = 10, B = 20, total complex: 4, 8, 12, 16, 20. C3$\overline{b}$ plotted. The lowest complex concentration yields no measurable C3$\overline{b}$.

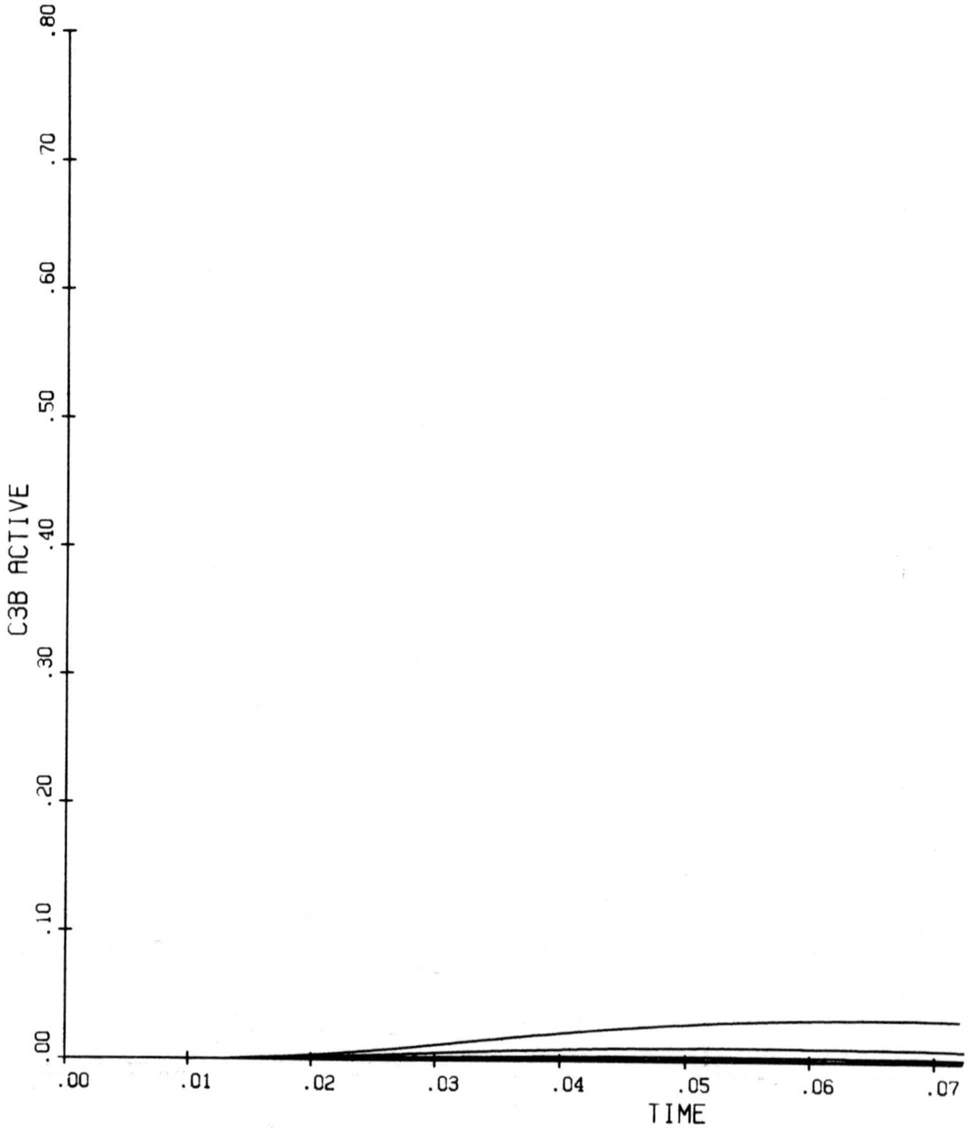

Fig. 10.5c. C2 = 1, B = 20, total complex: 4, 8, 12, 16, 20. $C3\overline{b}$ plotted. (Compare with 4d.)

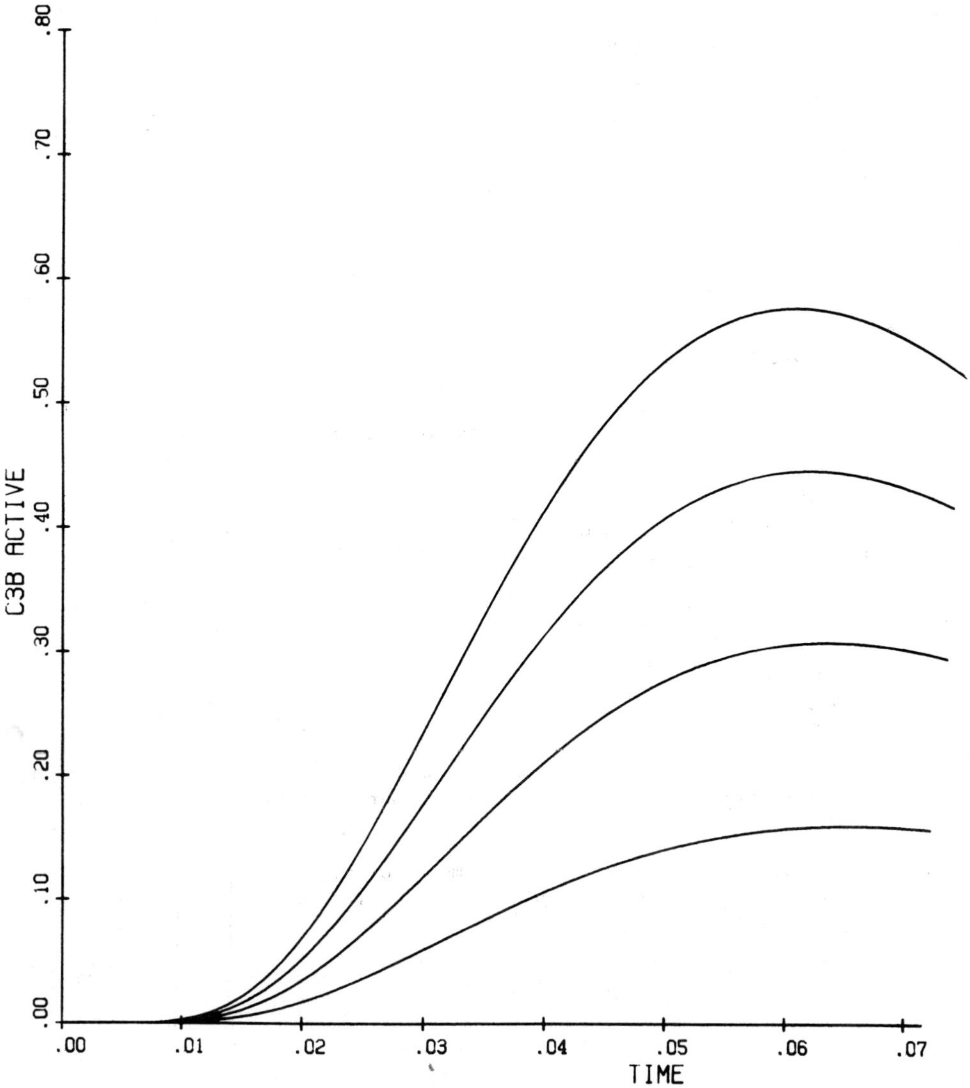

Fig. 10.6. Simulations illustrating the effect of varying C2 concentrations with both complex and C1 inhibitor fixed (at 10 and 20 units respectively) and C2: 5, 10, 15, 20. C3b plotted.

REFERENCES:

1. Banks, H. T., Modeling and Control in Biomedical Systems, Springer-Verlag, 1975.

2. Bell, G. I., Perelson, A. S. and Pimbley, G. H., editors, Theoretical Immunology, Marcel Dekker, 1978.

3. Benacerraf, B. and Unanue, E. R., Textbook of Immunology, Williams and Wilkins, 1979.

4. Colten, H. R., Complement synthesis, in Biological Amplification System in Immunology, N. K. Day, and R. A. Good, editors, Plenum, 1977, 47-67.

5. Day, N. K. and Good, R. A., editors, Biological Amplification Systems in Immunology, Plenum, 1977.

6. Donaldson, V. H. and Evans, R. R., A biochemical abnormality in hereditary angioneurotic edema: absence of serum inhibitor of C'1 esteras, Amer. J. Med. 35: 37-44 (1963).

7. Donaldson, V. H., Mertee, E., Rosen, F. S., Kretschmer, K. W. and Lepow, J. M., A polypeptide kinin in hereditary angioneurotic edema plasma; Role of complement in its formation, J. Lab. Clin. Med. 76: 986 (1970).

8. Gewurz, H. and Lint, T. F., Alternative modes and pathways of complement activation, in Biological Amplification Systems in Immunology, N. K. Day and R. A. Good, editors, Plenum, 1977, 17-45.

9. Griffin, F. M., Opsonization, in Biological Amplification Systems in Immunology, N. K. Day and R. A. Good, editors, Plenum, 1977, 85-113.

10. Merrill, S. J., Mathematical models of humoral immune response, in Modeling and Differential Equations in Biology, T. Burton, editor, Marcel Dekker, 1980, 13-50.

11. Merrill, S. J. and LeFever, A. V., A model of complement activation by antigen-antibody complexes, in Differential Equations, S. Ahmad, M. Keener and A. Lazar, editors, Academic Press, 1980, 221-233.

12. Merrill, S. J. and LeFever, A. V., The classical pathway of the complement system: a mathematical model, in preparation.

SOME INTRODUCTORY PHILOSOPHICAL REMARKS ON THE MATHEMATICAL MODELING OF COMPLEX SYSTEMS

Elliott W. Montroll
Institute for Fundamental Studies, Department of Physics and Astronomy, University of Rochester, Rochester, New York

INTRODUCTION

Since ancient times man has attempted to portray small portions of the complicated universe around him (sometimes even of himself) in some simplified representation which emphasizes one or several important features of the small portion which particularly fascinates him. Many styles of simplification or modeling have evolved over the centuries. Aesop wrote fables and the Bible quotes parables to model human frailties. The case histories by psychologists are models of complex human behavior patterns evidenced in subjects. The caveman's few strokes of wall markings are models that give us insight into the appearance of the animals he encountered. The Dutch and Flemish 17th century school of painting developed a remarkable two dimensional model of the detailed appearance of complex indoor and outdoor scenes. The charm of Picasso's work lies in the images evoked by the few lines in his pictorial models.

A university is a collection of modelers each following the traditional style of his academic discipline. The political scientists have their style in modeling political activities, the economists theirs in modeling the economy; the biologists' style of modeling characteristics of living organisms is somewhat different from the physicists' and chemists' style for inanimate matter.

The current availability of cheap computational facilities has steered the modelers in most fields toward a common course of increased dependence upon mathematical modeling. In this report I wish to present a brief survey of some mathematical models of complex systems, each chosen to emphasize a different style of modeling or to alert the listener (or reader) to difficulties which confront modelers and planners.

It is not sufficiently emphasized that the art of constructing reliable mathematical models is precisely the art of doing good science. Unfortunately, the modeler of policy matters does not enjoy the freedom of an academic scientist who often does good science by limiting the subject of his investigation. The academician may make a career of the hydrogen atom, the drosophila, or amorphous selenium. A modeler concerned with environmental topics, for example, must include subtle interactions between

a) population dynamics
b) natural resources
c) energy demands and forms
d) ecology
e) health
f) food and famines
g) economics
h) foreign trade, etc.

The demands upon his modeling efforts are enormous since the detailed understanding of these individual topics is as limited as that of the environment itself. Nevertheless, since important public policies may be based upon his results, he must apply the same severe critical analysis to his work as is common to the best academic researchers.

I consider Johannes Kepler to be the founder of the discipline of mathematical modeling as an aid to decision making. His three laws for the motion of the planets about the sun are the characterization of a magnificent model; and in his eyes more information about the planets meant better horoscopes. Kepler's official position as court mathematician and astrologer at Prague was prestigious even though his employer, Emperor Rudolph II, was generally in arrears in salary payments. In those days battles could not be fought nor royal marriages arranged without the advice of the court astrologer.

Since Kepler the astrologer was hired to make predictions, it is natural that Kepler the astronomer showed similar predelections. When he derived his orbit for the planet Mercury and prepared his "Rudolphine Tables" of future positions of the planet, he specifically predicted in 1627 that it would cross the solar disk on 7 November 1631, which it did.

In Kepler's court forecast of things to come in 1618, he predicted [1]

> "...that the month of May will bring great upheavals to the world." And, indeed, on 23 May 1618 two Prague councillors are thrown from the windows of the Town Hall: the 30 Years War has started. A favourable "coincidence" for Kepler's prognostic powers. And so, too, is the famous horoscope Kepler cast for his patron Wallenstein. From this 1608 horoscope Kepler reads "...Saturn damages the imagination and often produces fear," and that his subjects' "unusual nature is suited to important matters." But later--in 1624--he tells Wallenstein that March 1634 will entail a terrible confusion in the

country which will affect him. Wallenstein, in fact, was assassinated on 25 February 1634. Was the truth, as has been suggested, that the courageous Wallenstein of many battles died without defending himself--because he has been "living" his horoscope in constant fear and expectation of this very last predicted moment?

The latter conjecture is in accord with the beliefs of the time. There was a fatal syndrome called "planet," an affliction of those who became so depressed by an unfavorable horoscope that they confined themselves to their rooms, rejected food, became hysterical, and eventually died. John Graunt [7] (for summary, see also [13]), in his tabulation of causes of death in London in 1632, attributes 13 deaths out of a total of 9535 to planet. This number exceeded by 3 the 10 assigned to cancer and wolf but was less than the 62 "suddenly" (probably heart attacks).

It is remarkable that even with the tremendous intellectual machinery available today, the record of the validity of the predictions and recommendations of modern court counselors is not always significantly better than those of the astrologer of old.

The selection of the models discussed below reflects my own philosophy of modeling. In the early stages of the investigation of a process or system, the more complicated it be, the simpler should be its mathematical model. The first step in the scientific process is to isolate a small number of important variables or to aggregate many variables appropriately so that simple relationships may be found between the aggregates. The statistical style allows one to model the cumulative effect of many weak influences by the definition of an appropriate random force.

The complexity of a process or system does not necessarily depend on the number of relevant variables, but rather upon the number of dimensionless constants necessary to describe its state. The importance of dimensionless constants is emphasized in the first topic I wish to review, traditional scale modeling in fluid dynamics.

THE NAVIER-STOKES EQUATION AND THE USE OF DIMENSIONLESS CONSTANTS IN SCALING EXPERIMENTS IN HYDRODYNAMICS

The most fruitful modeling strategy for complex fluid dynamical systems is the use of dimensionless constants for design of experiments involving small scale physical models of full sized objects under investigation. The design of airplanes, ships, dams, harbors, canals, etc. would be impossible without scaling experiments. The physical basis and the experimental practice of this strategy will be described in terms of the Navier-Stokes equation for the flow field of an incompressible viscous fluid.

Let $v \equiv v(r,t)$ be the velocity of a fluid element located at r at time t of an incompressible fluid of density ρ and with kinematic viscosity ν. The Navier-Stokes equation is [16]

$$\frac{\partial v}{\partial t} + v\cdot\nabla v = -\nabla(p/\rho) + \nu\nabla^2 v + F/\rho, \tag{11.1a}$$

$p \equiv p(r,t)$ being the pressure on the fluid element at (r,t) and F the external force acting on the fluid. The equation of continuity (a consequence of the conservation of mass) for an incompressible fluid is

$$\nabla\cdot v = 0. \tag{11.1b}$$

Since Equation (11.1a) is a vector equation it corresponds to three scalar equations. These combined with (11.1b) yield four equations for the four variables, the pressure and the three components of the velocity vector. The force function will be specialized to be a gravitational force with

$$F/\rho = g. \tag{11.1c}$$

The Navier-Stokes equation is applicable to many processes involving fluids, such as the motion of ships and subsonic airplanes, and the flow of fluids through pipes. In the description of each of these processes one must introduce the boundary conditions required to specialize Equation (11.1) to the process being considered. The nonlinear term $v\cdot\nabla v$ in Equation (11.1a) is a tremendous mathematical nuisance. It makes hydrodynamics a hard subject. On the other hand, it contributes to making it physically a rich subject.

Since the state of the art of solving nonlinear partial differential equations has not become sufficiently advanced to solve Equation (11.1a) under the conditions required for the design of ships and airplanes or even for the prediction of the weather, it is important to develop experimental techniques for the investigation of the solutions of (11.1a) for engineering applications.

Scaling theory is based on the transformation of Equation (11.1) to an equivalent equation for dimensionless quantities. The velocity, the pressure, and even g in (11.1a) are all variables with dimensions--their numerical values depend on the units chosen. To obtain a dimensionless equivalent of (11.1) we measure the local velocity, pressure, etc. as a multiple of some important basic dimensions of the object responsible for the flow pattern being investigated. Suppose, for example, we wish to investigate the flow pattern of air around an airplane in flight. We let

V = average velocity of body being investigated

L = an important unit of length of the body (say the average width of an airplane wing)

P = average pressure in absence of body.

Then we can define a set of dimensionless quantities u', p', x', etc. by

$$v = Vv' \qquad x = Lx' \qquad p = Pp'. \tag{11.3a}$$

If we are concerned only with steady flow patterns, we can set $\partial v/\partial t = 0$ in Equation (11.1a). Now, if we set

$$\nabla' = i\partial/\partial x' + \ldots = L\nabla, \tag{11.3b}$$

Equation (11.1a) can be shown to have the form

$$v'\cdot\nabla v' = -P\nabla' p' + (1/R)\nabla'^2 v' + (1/F) \tag{11.4}$$

with dimensionless coefficients defined by

$$R = \text{Reynolds number} = VL/\nu \tag{11.5a}$$

$$P = \text{Pressure number} = P/\rho V^2 \tag{11.5b}$$

$$F = \text{Froude number} = V^2/Lg. \tag{11.5c}$$

There are certain flow regimes with the property that two out of the three terms on the right-hand-side of (11.4) can be neglected relative to the last remaining one. For example, suppose that only the term 1/F need be retained. Then the flow velocity field and engineering design parameters that depend on the flow field would be a function of only the dimensionless quantity F. Hence small scale model experiments could be made to obtain design data for the full scale engineering of the device of interest.

It is easy to show that for a 1000 ft ship operating in the 40 ft/sec range, with pressures being measured in units of atmospheric pressure (and noting that the kinematic coefficient of viscosity of water at 15°C is $\nu = 1.23 \times 10^{-5}$ ft^2/sec).

$$1/F = 20, \quad P = 0.69, \quad \text{and} \quad R^{-1} = 10^{-9}.$$

Hence, the 1/F term is the most important one on the right hand side of (11.4). Then, ship modeling can, to a first approximation, be based on Froude modeling; i.e., modeling with a dimensionless constant that depends on g.

The visual evidence of this effect appears in the bow wave produced by the ship. As the ship is propelled through the water it creates waves, their amplitude and wave length increasing with the speed of the ship. Thus, with increasing speed, a larger fraction of the energy supplied for propulsion is dissipated into the formation of the waves. This is exhibited in Figure 11.1 where we have plotted

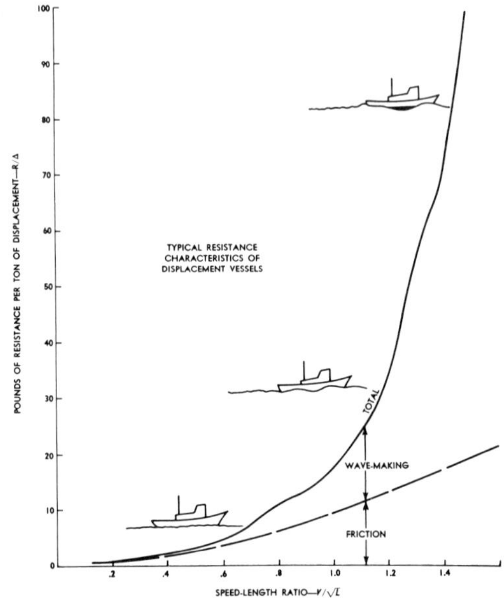

Fig. 11.1. Resistance versus $V/L^{\frac{1}{2}}$ in a characteristic Speed-Power curve for a displacement ship (note the ship-wave profiles), taken from [6].

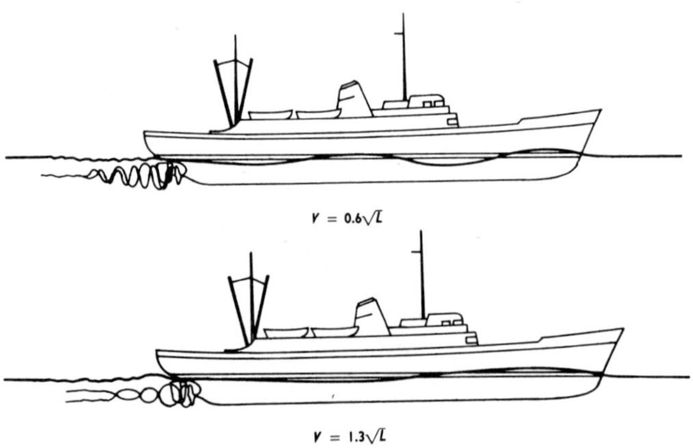

Fig. 11.2. Surface displacement ship wave-making signatures at various speeds, taken from [6].

experimentally determined (with models in a towing tank) pounds of resistance per ton of ship displacement as a function of Froude number [6]. Formulae exist for the conversion of resistance figures to effective horsepower for propulsion.

An important feature of Figure 11.1 is the rapid rise in resistance (and therefore fuel requirements) as the velocity (in knots) exceeds $1.3L^{\frac{1}{2}}$ (the length being in feet). This can be understood qualitatively [6] in terms of Figure 11.2.

At low velocities, the wavelength of the bow wave is short; with increasing velocity it lengthens. When the velocity (in knots) reaches $0.6L^{\frac{1}{2}}$ (again length in feet) there are about two and a half bow waves per hull length. Finally at the critical speed $V = 1.3L^{\frac{1}{2}}$ the wavelength of the bow wave becomes equal to the length of the hull [6]. Since the upper crest of the wave is aft of the bow and the lower crest fore of the stern, the ship is effectively going up hill when the critical speed is achieved. This tipped configuration thus requires considerably more propulsion power than a ship moving horizontally at low speeds. A 750 ft hull has the critical speed 35.6 knots; while for a 1000 ft one, $V_c \simeq 41$ knots. Therefore it becomes very expensive to operate a large ship whose cruising speed exceeds 35 to 40 knots.

The use of dimensionless constants need not be restricted to models in the physical sciences. I will demonstrate this by reviewing an excellent social modeling application due to R. J. Smeed [18] who considered the question: How far do people drive to work? Numerous surveys and questionnaires on this query have been analyzed in many cities.

Smeed argued that if there is a "law" or formula that would yield the answer to this question, it must involve a dimensionless constant derived from the average distance driven (a length) and some other characteristic length of a city. He chose the other length to be the square root of the area, A, of the city and by analyzing the available data arrived at the conclusion that

$$r = d/A^{\frac{1}{2}} \simeq 0.85. \tag{11.6}$$

While one city (Los Angeles) had an r value slightly larger than 1 and several had the ratio around 0.75, most had are values between 0.80 and 0.90. For example, in London 1960, r = 0.87.

Smeed then asked why r was about 0.85 and not of the order of 0.1 or 10. He satisfied his curiosity by considering several models, the first being a circular city with an internal regular grid of streets. He postulated workplaces to be uniformly distributed along the outer circle. Then he assumed that anyone who lived on the circle had an equal probability of working anywhere on the grid and found the numbers given in Figure 11.3. By averaging over all paths starting on the circle and ending on a grid of seven vertical lines and seven horizontal lines he found $d = 0.81A^{\frac{1}{2}}$, which is close to Equation (11.6). He made similar hypotheses about square and rectangular cities as well as models of Reading and Cambridge as indicated in the figure. The anomalous city, Los Angeles, is a long, narrow city.

THE TYRANNY OF MANY DIMENSIONLESS CONSTANTS

We have noted that simple questions about ships, airplanes, and flow of fluids through pipes can each, to a first approximation, be discussed in terms of a single

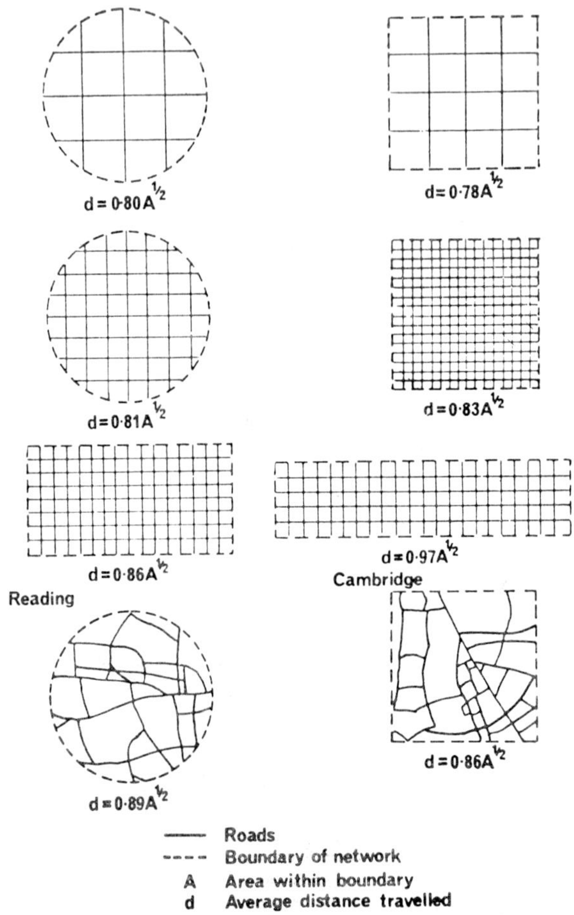

Fig. 11.3. Average distance traveled on some imaginary and real road networks [18].

dimensionless constant. Even without an understanding of hydrodynamics, ships sailed around the world; it was only with the advent of steam propulsion that naval architects sought to make calculations of the power required for a ship to perform in a desired manner. When sailors depended on the wind they tinkered with ship shapes and sail configurations to empirically improve performance.

It is remarkable that within ten years of the first flight of the Wright brothers, Igor Sikorsky (1913) built a four engine giant (92 ft wing span) capable of carrying a payload of 4½ tons. Sikorsky's Ilya Mourometz (1914) stayed airbourne for 6½ hours and carried six passengers [19]. It could have served as a prototype for a commercial airliner. However, World War I intervened and soon pilots were shooting at each other in aerial dogfights, and bombing cities of their opponents. By

1924, 21 years after Kitty Hawk the Imperial Airways flew Handley-Page Airliners in routes from Cairo to Africa and India [19]. The 1928 model H-P 42's had a passenger capacity of 38.

On December 2, 1942, Enrico Fermi and his colleagues produced the first sustained fission chain reaction, demonstrating the feasibility of a nuclear power plant. By fifteen years later, the first commercial nuclear power plant was operating.

The highly successful space program was initiated only a few decades after primitive rocket experiments. Robert Goddard's first success in shooting a rocket over a mile vertically was achieved on May 31, 1935; 26 years later the Soviet cosmonaut was the first human to encircle the globe in a rocket launched satellite.

The record of these successes has led the public, and even numerous scientists, to believe that with a little money and ingenuity, any desired scientific goal could be achieved. Unfortunately, this is not always true. Consider the magnetically confined fusion program which started as Project Sherwood in 1951. The aim of the project was to accelerate a plasma of ionized deuterons to a point where upon collision of each other, at an energy equivalent to a temperature of about 300×10^6°K, nuclear fusion would occur with a tremendous release of energy. Since 1/6500 of the hydrogen in ocean water is composed of deuterium it is considered by optimists that if the controlled fusion process can be successfully achieved, the energy problems of the world would be solved. Unfortunately, twenty-seven years and hundreds of millions of dollars later, energy by magnetically confined fusion seems even further away than it did in 1951. What has happened? Why has this branch of physics failed to live up to expectations.

I contend that the magnetically confined fusion program has fallen victim to the tyranny of many dimensionless constants. The great engineering successes of the past have involved processes which could, to a first approximation be characterized by a small number of dimensionless constants. Hence only a small number of model experiments were necessary to determine the feasibility of a project and to estimate the cost and difficulties to be surmounted. Even the space program was broken down into a number of subprojects, each of which could be analyzed in terms of a small number of dimensionless constants so that the results of many independent model tests could be used as a basis of the required full scale engineering designs.

The complication of the magnetically confined fusion program seems to be that all the hydrodynamic dimensionless constants (about eight) as well as several electromagnetic and nuclear dimensionless constants are intimately connected in the process of transforming a low density, low temperature plasma to a higher density very high temperature plasma. Since, as we shall now indicate, the cost involved in, or the time required for the understanding of the nature of a process characterized by N interacting dimensionless constants can be expected to grow exponentially with N, we should not be surprised with the slow progress in the field of magnetically confined fusion.

Let N be the number of dimensionless constants required to characterize a process. Then an experimental program might sample $n_1 \times n_2 \times \ldots \times n_N$ points in the N dimensional space of characterization. The cost of the program P should be proportional to the number of sampling tests; i.e.,

$$P = kn_1 \times n_2 \times \ldots \times n_N = k \exp N\left\{\frac{1}{N}\sum_{j=1}^{N} \log n_j\right\}.$$

Hence, if we define λ to be the average value of the logarithm of the number of observation for each dimensionless constant

$$P = k \exp N\lambda$$

as was suggested to be the case.

The genius of individual inventors has sometimes allowed one to cut costs and time by going directly to the correct regime of the dimensionless constant of interest without conducting model tests over a broad range of the dimensionless constants. We shall now indicate that on a purely probabilistic basis the likelihood of being able to cut corners by ingenuity decreases exponentially with the number of dimensionless constants.

Let a process variable be a function of a dimensionless independent variable x_1 of restricted range R_1. Now suppose that the process is operational when x_1 is in a subrange of R_1 which occupies a fraction f_1 of R_1. Then if one picks a value of x_1 at random the probability that the process is operative is f_1. Let us consider another process variable that is a function of two independent dimensionless variables x_1 and x_2, x_1 having a restricted range R_1 and x_2 having a restricted range R_2. Then, let us suppose that our required two variable process is operative when x_1 is in an appropriate subinterval of R_1, which occupies a fraction f_1 of R_1; and at the same time x_2 is in an appropriate subinterval of R_2, which occupies a fraction f_2 of R_2. If a value of x_1 and one of x_2 is selected at random the probability of our process being operative is $f_1\ f_2$, assuming that the process variables are independent.

Let us generalize to the case of a process variable being a function of n independent dimensionless constants $x_1, x_2, \ldots, x_n$, with the process being operative when each x_j is in a subrange of R_j which occupies a fraction f_j of R_j. Then the probability of the process being operative is

$$P_n = f_1 \dots f_n = \exp \sum_{j=1}^{n} \log f_j$$

$$= \exp(-n\mu)$$

where

$$\mu = \frac{1}{n} \sum_{j=1}^{n} \log(1/f_j).$$

Since each $0 < f_j < 1$, μ is a positive number and the probability of a "genius" selecting a set of points $(x_1 \dots x_n)$ which makes a process operative declines exponentially with the number of dimensionless constants n.

Certain social situations and environmental processes might also depend on a large number of dimensionless constants. The understanding of these processes is not exempt from the tyranny of many dimensionless constants; nor is an attempt to make policies exempt, without a considerable insight into the manner that a change in a single dimensionless constant influences others. Just as the enthusiast for magnetically confined nuclear fusion knows how he would like to solve the energy problem so the enthusiast for social and environmental reform knows how he would like to make our lives full of harmony and beauty. Unfortunately both of these classes of enthusiasts remain dreamers until the tyranny of many dimensionless constants is overcome.

THE VERHULST POPULATION GROWTH MODEL AND A NOTE OF CAUTION ON THE CONVERSION OF DIFFERENTIAL EQUATIONS TO DIFFERENCE EQUATIONS

The pessimist's equation for population increase is the Malthus equation. It states that the population, n(t), is determined by

$$dn/dt = kn. \tag{11.7a}$$

Its solution is the exponential function

$$n(t) = n(0)\exp kt. \tag{11.7b}$$

Verhulst was one of the few mathematicians in Belgium at the birth of its independence (1830). Since Belgium was a small country with a high population density he preferred the more optimistic model equation for population growth [20]

$$dn/dt = kn(1 - \frac{n}{\theta}) \tag{11.8}$$

which yields a saturation level of population θ. The solution of (11.8) is easily shown to be

$$n(t) = \theta n(0)\{n(0) + [\theta - n(0)]\exp - kt\}^{-1}. \tag{11.9}$$

If we let x(t) be the fraction of the way to saturation:

$$x(t) = n(t)/\theta, \qquad x_0 \equiv n(0)/\theta. \tag{11.10}$$

Then

$$x(t) = x_0\{x_0 + (1 - x_0)\exp - kt\}^{-1}. \tag{11.12}$$

This equation is equivalent to

$$\log \frac{x(t)}{1 - x(t)} = \log \frac{x_0}{1 - x_0} + kt. \tag{11.13}$$

The population growth of many countries can be fitted rather well to Equation (11.9), which is often called the logistic model. It was rediscovered by Pearl [15] and Reed and much studied by them in the 1920's. A typical fit is shown in Figure 11.4 where (11.9) is compared to U.S. census data. Notice that the model is remarkably good until the depression of the 1930's when the growth rate declined from its expected value. After World War II the baby boom more than compensated for the depression decline. While there are, of course, more sophisticated ways to make population projections, Equation (11.9) is still interesting.

Probably the most important use of the Verhulst equation in the future will be for the estimation of the manner in which new products or technologies will replace old ones. This application was proposed by J. Fisher and R. Pry [4], who investigated replacement dynamics in more than twenty industries. In their model, x is to be interpreted as the fraction of production (market) by a new technique (product) and (1 - x) the fraction remaining by (for) the old. Some of the replacement examples cited to follow (11.12) and (11.13) were the replacement of the Bessemer converter method of steel production by the open hearth method, soap by detergents, wood-based floors in new construction by plastic, and the replacement of oil-based paints by water-based. Replacement data plotted as the time variation of x/(1 - x) on semi log graph paper yielded excellent straight lines in accordance with (11.13). On

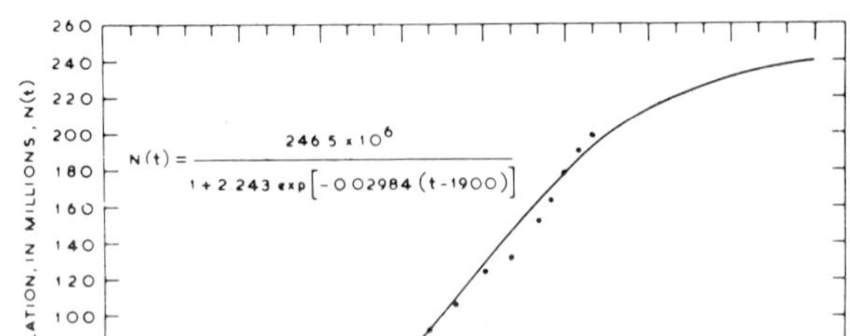

Fig. 11.4. Population growth of the U.S. fitted to the Verhulst equation so that the census data of 1840, 1900, and 1960 fit the curve exactly [13].

this basis one might consider some adaptation of (11.8), say

$$d \log x/dt = k(1 - x) \tag{11.8a}$$

to be the canonical differential equation of the replacement process. One might also consider the case of the existence of some extenuating circumstance (or "force") accelerating or deterring the rate of replacement. Such a force F could be recorded as a perturbation of (11.8a) according to

$$d \log x/dt = k(1 - x) + F. \tag{11.8b}$$

We now display two examples of replacement processes which follow (11.8a) for a time and then become affected by the influence of some important forces.

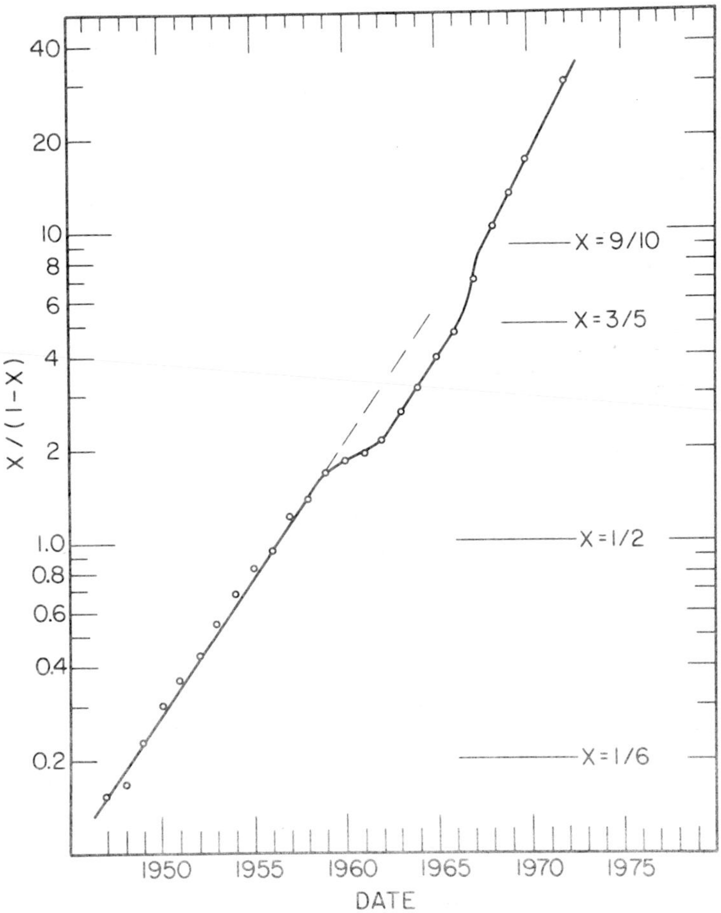

Fig. 11.5. Replacement of rail travel by air travel. The variable x is the ratio of annual passenger miles by air to total air plus rail passenger miles in continental U.S.

F. Harbus and the author have examined the replacement of railroad passenger service by air transport in intercity travel. The required data for the period 1946-1970 was taken from "Historical Statistics of the United States" (Bureau of Census, 1975 edition), tables Q 309-310 for annual rail passenger miles (table Q 308 being omitted as representing local commuter travel) and Q 585 for annual air passenger miles. Figure 11.5 is a graph of the ratio of air passenger miles to rail passenger miles plotted on semi log paper.

Notice that in the period 1947-1959 the data follows a straight line in accordance with Equation 11.13. The deterrence in replacement evolution evident in 1960-61 seems to have been due to several unusually long labor strikes of airline workers and to the public's response to a series of serious airplane accidents. By 1962 the system recovered with the replacement curves continuing with their odd slope until the late 1960's. Then accelerated replacement to the end of the decade occurred at the time the largest rail passenger carrier, the Penn-Central line, was suffering through its pre-bankruptcy and bankruptcy condition. The management of passenger rail service was reorganized in the 1970's by AMTRAC.

The role of intercity busses was omitted from the above discussion since, during the period investigated, the fraction of the total passenger traffic maintained by them changed very little.

It occurred to Robert Herman and the author [11] that the industrial revolution of the 19th century was an evolutionary process converting the labor force of a country from its agricultural nature to an industrially directed one in a manner that might be described by the logistic equation model. The fraction of the labor force doing agricultural work declined as the fraction involved in industry grew.

Prior to 1840 the ratio of nonagricultural workers to agricultural workers in the labor force of the U.S. remained fairly constant over many decades. The ratio of the fraction of nonagricultural workers to agricultural workers in the U.S. is plotted as a function of time in Figure 11.6 (as is the corresponding fraction for Sweden). The U.S. data was obtained from Statistical Abstracts of the United States. It is remarkable how well the data fits the straight line defined by the logistic equation for a period of about 100 years.

The rate at which agricultural workers left the farms in the early 1940's was somewhat greater than would have been expected on the basis of the logistic equation. The new force was generally a telegram from President Roosevelt which started with the word "Greetings...". After World War II many of the young men who responded to the Greetings did not return to the farm. The curve for $f/(1 - f)$ (nonagricultural to agricultural worker ratio) for Sweden follows that of the U.S. rather

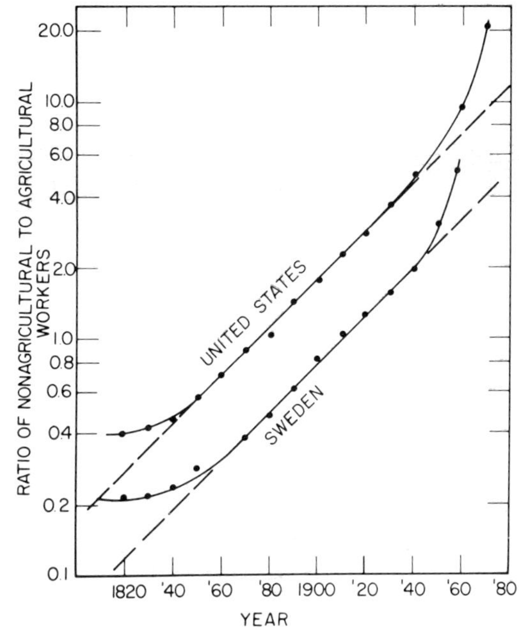

Fig. 11.6. Variation of the ratio of nonagricultural to agricultural labor force [2] in U.S. and Sweden, 1820-1970.

well. While the young Swedish farm hand did not get a greetings telegram, he was enticed to move to the city in response to the higher wages paid by companies which were selling factory products to pleading customers around the world.

The evolution of social and industrial patterns might be considered as a sequence of replacements of old ideas, old habits, old mores, old products and old practices by new ones. Hence as one attempts to make more detailed models of the evolutionary process, he would expect the Verhulst equation to continue to appear prominently.

Most rate equation models for complex systems couple various important elements of the system. With modern computers it is not unusual to have coupled nonlinear rate equations in a dozen or more variables. A complication of this style modeling is the enormous number of rate constants required. Even in a linear system of a dozen coupled variables 144 rate constants must be assigned. Considerable boldness is required to have the confidence that so many parameters can be determined with sufficient accuracy to be able to depend on the qualitative features of numerical solutions of the coupled rate equations.

The remainder of this section is concerned with still another type of problem that is not always recognized. It will be shown that in the process of converting a nonlinear differential equation to a difference equation, solutions exist to the difference equation which are qualitatively different from those of the differential equation. As our example we use the Verhulst model in the variable x defined by (11.10).

$$dx/dt = x(1 - x) \quad \text{with} \quad x > 0, \quad t > 0. \tag{11.14}$$

From Equation 11.12 we note that there are two classes of solutions of this equation. If (a) $x_0 > 1$, then the solution decays monotonically to $x = 1$ as t increases from zero, and if (b) $0 < x_0 < 1$, the solution increases monotonically to $x = 1$ as t increases. There are no periodic solutions and, if $x_0 > 0$, $x(t)$ never vanishes or becomes negative for $t > 0$.

If (11.14) were to be solved numerically, it would be natural to replace it by the difference equation

$$x_{n+1} = x_n + \tau x_n(1 - x_n) \quad \text{with} \quad t = n\tau \tag{11.15}$$

$$= \tau x_n[(1 + \tau^{-1}) - x_n]. \tag{11.16}$$

We seek the qualitative characteristics of the solution of this difference equation when

$$0 < x_0, \qquad 0 < \tau < 1 \tag{11.17}$$

in four regimes of initial values

$$\text{(a)} \quad 0 < x_0 < 1 \qquad \text{(b)} \quad 1 < x_0 < \tau^{-1} \tag{11.18a}$$

$$\text{(c)} \quad \tau^{-1} < x_0 < 1 + \tau^{-1} \qquad \text{(d)} \quad 1 + \tau^{-1} < x_0. \tag{11.18b}$$

The character of the solution of (11.14) in these regimes can be developed from the following properties of the mapping

$$y = x[1 + \tau(1 - x)] \tag{I}$$

$$= \tau x[(1 + \tau^{-1}) - x]. \tag{II}$$

$$\text{with } 0 < \tau < 1: \tag{III}$$

i) If $0 < x < 1$, then $y > x$. This follows from I, since

$$1 + \tau(1 - x) > 1.$$

ii) If $x < 0$, then $y < x$. This follows from

$$1 + \tau(1 - x) > 1 \quad \text{if} \quad x < 0$$

so that y is negative (from I) when x is negative. Then

$$|y| = |x||1 + \tau(1 - x)| > |x|$$

as required.

iii) $dy/dx > 0$ for all $0 < x < 1$, so that y is an increasing function of x in that range. To prove this statement we first note that y is a continuous function of x and that

$$dy/dx = (1 - \tau x) + \tau(1 - x).$$

Then $dy/dx > 0$ since $(1 - x) > 0$ and $\tau x < 1$. Since $y(1) = 1$ this also implies

iv) If $0 < x < 1$, then $y < 1$.

v) If $x > 1$, then $y < x$. This follows from I and

$$\tau - \tau x < 0 \quad \text{or} \quad 1 + \tau(1 - x) < 1.$$

vi) If $1 < x < \tau^{-1}$, then $y > 1$. For proof we note first that

$$y(1) = 1 \quad \text{and} \quad y(\tau^{-1}) = 1.$$

At the midpoint of the x interval $x = \frac{1}{2}(1 + \tau^{-1}) = (1 + \tau)/2\tau$.

$$y([1 + \tau]/2\tau) = (\tfrac{1}{2} + \tfrac{1}{2}\tau^{-1})^2 > 1,$$

remembering the previous section. Inasmuch as $y(x)$ is a continuous quadratic function of x it cannot achieve the value 1 at any other choice of x than $x = 1$, and τ^{-1}. Therefore, $y > 1$ if $1 < x < \tau^{-1}$ as required.

We are now in a position to find x_n for various initial regimes of x_0:

a) If $0 < x_0 < 1$, then from (i) and (ii), by letting $x \equiv x_n$ and $y \equiv x_{n+1}$

$$x_0 < x_1 < x_2 < \ldots < 1$$

a monotone increasing function of n which never exceeds 1.

b) If $1 < x_0 < \tau^{-1}$, then from (v) and (vi)

$$x_0 > x_1 > x_2 > \ldots > 1,$$

a monotone decreasing function of n never becoming < 1.

In the above two regimes the solutions of the difference equation (11.15) have the same character as those of the differential equation (11.14).

Now consider the remaining cases:

c) Let $\tau^{-1} < x_0 \leq 1 + \tau^{-1}$. We may then write

$$x_0 = \tau^{-1} + \varepsilon \quad \text{with} \quad 0 < \varepsilon \leq 1.$$

Hence

$$x_1 = (1 - \varepsilon)(\tau\varepsilon + 1) = 1 - (1 - \tau)\varepsilon - \tau\varepsilon^2 < 1.$$

The smallest value achievable for x_0 in the present regime corresponds to $\varepsilon = 1$ at which point $x_1 = 0$. Hence, if $\tau^{-1} < x_0 < 1 + \tau^{-1}$ then

$$0 < x_1 < 1.$$

Combining the above inequalities with i) we find

$$0 < x_1 < 1 < x_0 \quad \text{and} \quad x_1 < x_2 < x_3 < \ldots < 1.$$

Therefore, in the present regime x_n is not a monotone function of n since it first decreases with increasing n and later increases with increasing n unlike the solutions of the differential equation.

d) Let $x_0 > 1 + \tau^{-1}$. Then $\tau(x_0 - 1) > 1$ or

$$\tau(1 - x_0) < -1$$

so that

$$x_1 = x_0[1 + \tau(1 - x_0)] < x_0[1 - 1] = 0.$$

Then, from (ii)

$$\ldots < x_3 < x_2 < x_1 < 0$$

all x_n for $n \geq 1$ are negative. Hence, when $x_0 > 1 + \tau^{-1}$, x_n eventually becomes and remains negative for succeeding n, yielding a form of solution of the difference Equation (11.15) which is not obtained from the Verhulst differential equation.

The conclusion to be drawn from the above analysis is that there are regimes of initial conditions which yield solutions of difference equations which are qualitatively different from the classes of solutions of the differential equations that they were intended to mimic. Incidentally, several authors [17] have studied (11.15) for $\tau \geq 1$ and have found it to have periodic solutions for some special initial conditions, another class not extant in the solutions of (11.14). There is no reason why the difference equations derived from coupled sets of nonlinear differential equations should not have classes of solutions that are not obtained from the differential equations.

There is another nonlinear difference equation which I claim to be the more appropriate representation of (11.14) than (11.15) for the regime $x > 0$, and $t \geq 0$ in that its solutions completely mimic the solutions of (11.7). The more appropriate alternative equation is, for $0 < \tau < 1$

$$x_{n+1} = x_n/[1 - \tau(1 - x_n)]. \tag{11.19}$$

It might be argued that if the right hand side of (11.19) is expanded in powers of τ and only the term of $0(\tau)$ is retained, then (11.19) would become equivalent to (11.16). This is true only for certain regimes of initial conditions and corresponding values of τ.

Equation (11.19) has the charm that it is easily solved in closed form by letting $y_n = 1/x_n$. Then (11.19) becomes linear

$$y_{n+1} = y_n(1 - \tau) + \tau$$

or

$$(y_{n+1} - 1) = (y_n - 1)(1 - \tau)$$

so that

$$y_n = 1 + (y_0 - 1)(1 - \tau)^n$$

and

$$x_n = x_0\{x_0 + (1 - x_0)(1 - \tau)^n\}^{-1} \tag{11.20}$$

which has the same qualitative monotonicity properties as x(t) defined by (11.12), if $0 < \tau < 1$. Since $t = n\tau$, n does not have to be very large to make the approximation

$$(1 - \tau)^n = (1 - \frac{t}{n})^n \sim e^{-t}$$

as required to pass from (11.20) to (11.12) as $\tau \to 0$. When $\tau = 2$, (11.19) has solutions of period 2 as is evident from (11.20).

ON SINGLE LANE TRAFFIC AND POTENTIAL INSTABILITIES IN SYSTEMS WITH TIME LAGS

Human relationships in pairs, informal groups, and in organizations are composed of a sequence of stimuli and responses. The stability of relation patterns is important in all phases of life. There are, of course, many examples of instabilities that arise from a small stimulus of one party being strongly amplified through the response of another with continuing amplification as the response acts as a stimulus to a next person or is fed back to the initiator of the pattern.

One of the causes of instability is the inherent time lag between stimulus and response. It will be indicated below, through an example that when the stimulus is oscillatory in nature instabilities may develop when too strong a response is activated by the stimulus if the response time lag is large.

There are generally three contributions to the total time lag between stimulus and response:

i) perception time to recognize nature of stimulus
ii) decision time required to choose response
iii) execution time for response.

All personal interaction processes from driving in a line of traffic, to industrial planning to international peace negotiations involve these three lag mechanisms.

An interesting human stimulus response relationship that exhibits the ideas discussed above is the interaction of a set of drivers in single lane traffic. Several of us investigated this follow the leader type traffic behavior a number of years ago and have developed a mathematical model to describe the resulting traffic flow ([2], [5], [12], [14], [3]).

Consider a line of identical cars moving with constant velocity v with a space a allotted to each car (a being the distance between the front bumper of the j-th car and that of the (j + 1)-st for all j). In our description of a line of "real" traffic we employ a coordinate system moving with velocity v to the right and let $x_j(t)$ be the deviation of the position of the j-th car at time t from where it would have been in a perfectly moving line. We also let $v_j(t)$ be the deviation in velocity of the j-th car from its ideal value.

A set of models of increasing complexity has been developed by a combination of car following experiments and stability arguments. The simplest model corresponds to a rather high density situation with each driver-vehicle combination suffering but slight deviations from an ideal flow pattern. The following stimulus response equation ([2], [12])

$$dv_j/dt = \lambda(v_{j-1} - v_j) \qquad j = 0, 1, 2, \ldots \tag{11.21}$$

with λ being the response coefficient yields a stable flow pattern in that any fluctuations in the motion of the lead car is diminished as the response to that fluctuation by successive drivers propagates down the line of traffic. Equation (11.21) is a statement that the j-th driver accelerates or decelerates by an amount proportional to his velocity relative to that of the (j - 1)-st car.

The stability of the model (11.21) is analyzed by postulating that the lead car drives in an oscillatory speed pattern of frequency ω, so that

$$v_0 = f_0 \exp i\omega t. \tag{11.22}$$

Then the j-th car responds with an amplitude f_j so that

$$v_j = f_j \exp i\omega t \tag{11.23}$$

where

$$i\omega f_j = \lambda(f_{j-1} - f_j). \tag{11.24}$$

By solving for f_{j-1} and iterating

$$f_j = f_0/(1 + i\omega\lambda^{-1})^j. \tag{11.25}$$

Hence the amplitude of the oscillatory response of the j-th car is

$$|f_j| = f_0/(1 + \omega^2\lambda^{-2})^{n/2} \tag{11.26}$$

which damps out, approaching zero as $n \to \infty$ as is required if the line of traffic is not to act as an amplifier of small disturbances. One finds that

$$v_j(t) = f_0(1 + \omega^2\lambda^{-2})^{-\frac{1}{2}n} \exp i[\omega t - j \cos^{-1}(1 + \omega^2\lambda^{-2})^{\frac{1}{2}}] \tag{11.27}$$

for all $\omega > 0$. Hence the velocity of propagation of a fluctuation down a line of traffic in this model is

$$\text{velocity of propagation} = \omega/[\cos^{-1}(1 + \omega^2\lambda^{-2})^{\frac{1}{2}}] \tag{11.28}$$

in units of cars per second.

The next more complicated model investigated is still linear but with the incorporation of the time lags outlined at the beginning of this section. Let Δ be the total time between the appearance of a fluctuation in the motion of vehicle $j - 1$ and the response of vehicle j. Then the generalization of (11.21) is

$$\frac{dv_j(t + \Delta)}{dt} = \lambda[v_{j-1}(t) - v_j(t)]. \tag{11.29}$$

Proceeding in the manner described in the derivation of Equation (11.27) we find its generalization to be

$$\begin{aligned} v_j(t) &= (1 - \omega^2\lambda^{-2} - 2\omega\lambda^{-1}\sin \Delta\omega)^{-\frac{1}{2}j} \\ &\quad \times \exp i\{\omega t - j \cos^{-1}[1 + \omega^2\lambda^{-2} - 2\omega\lambda^{-1} \sin \Delta\omega]\}. \end{aligned} \tag{11.30}$$

The amplitude factor decreases with increasing j if

$$1 + \omega^2\lambda^{-2} - 2\omega\lambda^{-1} \sin \Delta\omega > 1$$

or if

$$\omega/\lambda > 2 \sin \Delta\omega. \tag{11.31}$$

Low frequencies present the greatest limitation to the sensitivity λ. In the low frequency limit $\sin \Delta\omega \simeq \Delta\omega$. Hence for $|v_j(t)| \to 0$ as j increases we must have

$$\lambda\Delta < \frac{1}{2} . \tag{11.32}$$

When the inequality is violated $|v_j(t)|$ increases with j and the line of traffic acts as an amplifier of small low frequency disturbances.

A stability condition such as (11.32) is expected because if the time lag between an input fluctuation and a response is long and the response is very strong, one may be responding too strongly to a situation that has passed. The response then might be in phase with the fluctuation, further deteriorating the control. Equation (11.32) suggests that stability is best achieved by making small responses to small fluctuations soon after they occur. This is a good policy for the stabilization of social, economic and political interactions as well as those common to traffic flow. It is somewhat contrary to the manner that politicians build reputations--by responding dramatically to crises that have long been simmering. The response is not always a stabilizing one.

One of the most serious political mismatches between time lags and strength of response was the Austro-Hungarian response to the 28 June 1914 assasination of Archduke Ferdinand. After nothing was done for a month, on 23 July, an ultimatum was sent to the Serbians giving them twenty-four hours to make a suitable response, which they could not have done and the process of world destabilization started. It must be admitted, however, that some of the protagonists were not seeking stability.

An important feature missing from the model equations (11.29) is a natural dependence of the response parameter λ upon car spacing. Clearly a motorist responds more strongly to a car directly in front than to one that is far ahead. In the final model to be discussed it is postulated the λ is inversely proportional to the local space per car in the line of traffic [5]:

$$\ddot{y}_j(t + \Delta) = \lambda_0 \left\{ \frac{\dot{y}_{j-1}(t) - \dot{y}_j(t)}{y_{j-1}(t) - y_j(t)} \right\} \tag{11.33}$$

where $y_j(t)$ is now to be interpreted as the position of the j-th car at time t. This law fits experimental data taken from isolated test tracks as well as that observed in city driving over a large range of speeds and traffic densities.

It is easy to integrate these equations and average over a long line of cars to obtain an equation of state for traffic that relates flow rate, (say in cars per hour passing a given point), to traffic density (cars/mile). Clearly, if v is the mean speed of the traffic

$$q = v\rho = \text{flow rate.}$$

The density in cars per mile is the reciprocal of the spacing in miles per car (whose local value is $y_{j-1} - y_j$). In the derivation of the following result it is assumed that the lag Δ (of order 1.5 seconds) can be neglected compared with the time t used in the calculation of averages (the line of traffic being postulated to be stable). The equation of state derivable from (11.33) by integrating and averaging is ([5], [14])

$$q = v\rho = \lambda_0 \rho \log(\rho_c/\rho) \tag{11.34}$$

where v is the arithmetic mean of velocity averaged over many cars and ρ is the geometric mean of the density. The quantity ρ_c is the close packing density for cars bumper to bumper. In this state the flow rate of a line of traffic is zero as all drivers have noticed. Equation (11.34) has been verified in a number of experiments and observations ([3], [8]).

ON THE STATISTICS OF EXTREME VALUES

Small scale model testing is great and rate equation or bookkeeping type box models can be very helpful, but some systems are too complicated to be treated by these methods. Even when they are applicable, the required detailed analysis might not be the most efficient method of obtaining desired information. The statistical style offers an alternative way of investigating and monitoring complex systems.

I wish to close my discussion with a brief review of a branch of statistics, extreme value statistics, that is not widely known to decision makers and applied scientists. This subject is concerned with the question of how long one might have to wait for some variable to achieve an extreme value that has not yet been observed. It also gives a scheme to monitor random processes to determine when their nature is changed. Suppose, for example, it is predicted that a random variable should not reach some extreme level for five hundred years. Then, if it is observed to have achieved that level in five years the anomaly might be interpreted to indicate that the random process which generates the fluctuation in the variable may have changed its character.

I start the formal presentation of extreme value statistics by considering the "planners dilemma" involved in the planning of protection for some extreme events, say floods, earthquakes, financial reverses, etc. While the probability of the extreme event diminishes rapidly as a function of its magnitude (see Figure 11.7) the unpleasantness it might bring grows rapidly with its magnitude as does the cost

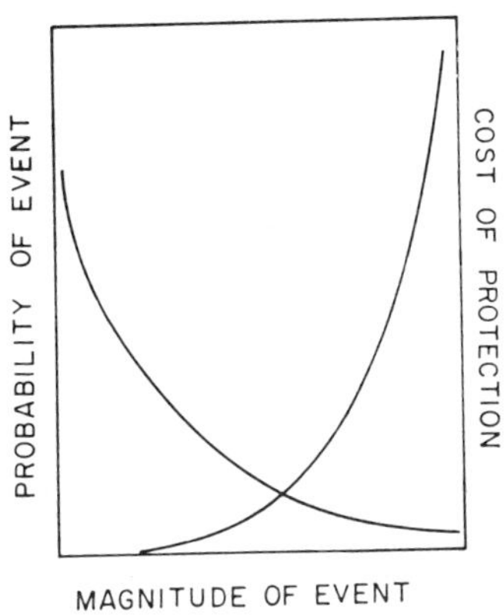

Fig. 11.7. The Planners Dilemma.

of protection against it. The planner is presented with the need to spend increasingly larger amounts of money to prepare for events which become decreasingly likely. A strategy we might consider would be to (a) estimate the expected lifetime of the device to be protected (or of the device affording the protection), (b) project the magnitude of the expected extreme event in that period, (c) introduce a safety factor α (> 1), and prepare for the expected event of great magnitude in the period αt. Of course, this strategy is applicable only if one can reasonably estimate the magnitude of the expected extreme events. It is for this purpose that a statistical model pioneered by E. T. Gumbel ([9], [10]) has been developed. Our discussion below will be given in terms of flood levels of rivers. The generalization to other situations will be obvious.

Every day the maximum river height at some location might be recorded. Over a number of years the mean daily height could be calculated as well as a distribution function of the deviation of the observed daily maxima from that mean. Now, let us suppose that river levels are generated by a stationary stochastic process whose character depends on winds, rainfall, temperature, melting rate of snow, etc., in a way which, while not understood, does not change from year to year.

The Gumbel theory of extreme values is concerned with the distribution of annual maxima, those levels which are the largest observed each year; i.e., the maximum value of daily maxima recorded in a single year. It can be shown that if (a) the basic stochastic process for the generation of levels is stationary, (b) the distribution of <u>daily</u> maxima levels is of exponential type (a class of distributions of which exponential, normal, log normal and Poisson are members), (c) correlations between fluctuations in daily maxima die out in a period much less than one year, then the cumulative distribution of annual maxima has the form

$$\phi(x) = \exp - (\exp - y) \tag{11.35a}$$

$$\text{with } y = \alpha(x - u). \tag{11.35b}$$

Here u is the mean annual extreme value and α is a parameter whose value can be deduced from empirical data.

The derivation of (11.35) is available in references [9], [10], and [14]. It is somewhat lengthy but not difficult. We derive (11.35) here for the special case of an exponential daily cumulative distribution function

$$F(x) = 1 - e^{-\alpha x}, \text{ when } x \geq 0. \tag{11.36}$$

We divide the time interval whose rare event statistics we wish to analyze into n equal subintervals (say n days) with F(x) being the cumulative distribution function of a variable x in each subinterval. If we choose the basic time interval to be one year, then n = 365. Let us postulate the observations in each time interval to be mutually independent. Then the probability that the largest value of x observed in n subintervals is less than x is

$$\phi_n(x) = [F(x)]^n, \tag{11.37}$$

since for each independent subinterval it must be less than x. Then from (11.36)

$$\phi_n(x) = (1 - e^{-\alpha x})^n. \tag{11.38}$$

Now define a number u_n such that

$$n \exp(-\alpha u_n) = 1. \tag{11.39}$$

Then

$$\phi_n(x) = \{1 - \frac{1}{n} \exp - \alpha(x - u_n)\}^n \tag{11.40a}$$

$$\sim \exp(-e^{-y}) \text{ for large n} \tag{11.40b}$$

where

$$y = \alpha(x - u_n). \tag{11.41}$$

The significance of u_n is determined by maximizing the probability density function

$$F_n(x) = d\phi_n/dx \sim \alpha \exp - (y + e^{-y}).$$

Then it is easy to show that $x = u_n$ is the value of x which is the most likely extreme value found in a set of n observations, since $F_n'(u_n) = 0$. Generally the parameter α is precisely

$$\alpha \equiv \alpha_n \equiv nf(u_n) \tag{11.42}$$

where f(x) is the probability density function of the daily distribution

$$f(x) \equiv dF/dx.$$

Gumbel constructed a special graph paper such that a plot of cumulative distribution functions of the form (11.40) appears as a straight line. The average number of years, T(x), which one has to wait for a given value of the annual maximum level X to exceed a value x is

$$T(x) = 1/[1 - \Phi(x)]. \tag{11.43}$$

The Gumbel distribution of extreme values has been applied to river levels, snow levels and other weather phenomenon, to strength of materials and to certain aspects of vital statistics. To demonstrate the use of the theory in making predictions, we consider death statistics records [9] in Switzerland in the period 1879-1933. As shown in the following table, in one year out of the 55 examined, the oldest person who died was 97, in three out of the 55 the oldest who died was 98.

Oldest ages in Switzerland, 1879-1933, both sexes

Oldest age	Number of individuals	Cumulative number, m	Plotting positions, m/(N + 1)
97	1	1	0.018
98	3	2 to 4	0.036 to 0.071
99	6	5 to 10	.089 to .178
100	14	11 to 24	.196 to .428
101	12	25 to 36	.446 to .643
102	7	37 to 43	.660 to .768
103	7	44 to 50	.786 to .893
104	3	51 to 53	.911 to .946
105	1	54	0.9643
106	1	55	.9821
	N = 55		

The most common age at which the oldest person died in a given year was 100. For reasons discussed in reference [10] Gumbel recommends that the plotting position for the cumulative number n be at m/(N + 1). On this basis, the cumulative distribution of the oldest ages at death in Switzerland is plotted in Figure 11.8 on Gumbel graph paper. A given value of $\phi(x)$ listed as frequency at the bottom of the graph paper is associated with a particular "return period" listed at the top as obtained from (11.43). Thus, for example, if one asks how many years will be required on the average for the oldest person to die in a given year in Switzerland to be older than 105, the graph would indicate 20 years. The fact that a good straight line can be drawn through the recorded cumulative distribution function substantiates the idea that the data is consistent with the Gumbel distribution. Many similar analyses are given in references [9] and [10].

Now suppose that after a relatively few years the oldest age at death in Switzerland is 112 or greater. This might be interpreted in either of two ways; first, that a fantastically unlikely event occurred, since on the average waiting time should have been about 100 years for the occurrence. Second, the stochastic processes which determine age at death for old people has changed its character. This might eventually be the case when a new understanding of the aging process, or cancer causes, or of circulatory ailments is achieved and with it a new mode of caring for the aged which leads to increased life span for centegenarians. At least the appearance of points completely off the Gumbel trend line calls one's attention to the possibility of a basic change in the process and can act as an indicator. Thus Gumbel graphs might be used to monitor processes, to judge when and if observations change character.

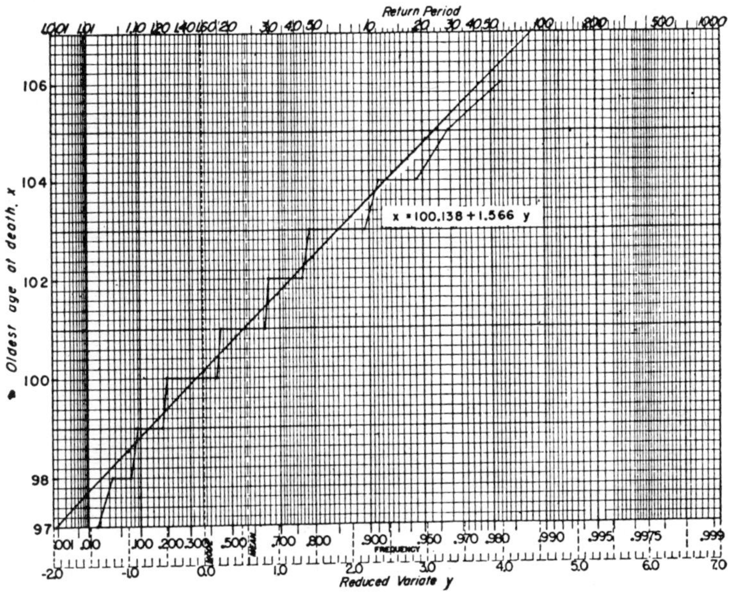

Fig. 11.8. Oldest ages at death, Switzerland, 1879-1933. From reference [9].

Gumbel has noted that such unexpected observations can also serve as economic indicators when economic data is plotted. Variations in the Dow-Jones industrial averages are often used to exhibit trends in the stock market. These daily averages reflect both the stochastic, or random, component of the market, and any systematic trends which might be developing. A trend line might merely reflect inflation and not necessarily a fundamental change in the relation of the public to the market. The theory of extreme values discussed above is applicable only to the random component of the market. Hence, if it is to be applied to boom-watching, the trend component must be subtracted from Dow-Jones averages.

The Dow-Jones central value characterizes the trend component. It is defined as the arithmetic mean of the median of daily highs and the median of the daily lows over a calendar year. An annual maxima can be defined as the difference between the largest daily high and the annual central value. These annual maxima for the years 1897-1947 have been plotted in Figure 11.9. Even without resorting to extreme value theory, it is clear that the stochastic process associated with the period 1928-1932 is quite unusual. The annual maxima were listed according to size and their cumulative distribution function was plotted by Gumbel on extreme value paper. The results are given in Figure 11.10. During the period 1897-1927 the straight line fitted the data quite well while the point representing 1928 is almost out of the picture, with its return period being over 2000 years. This strongly

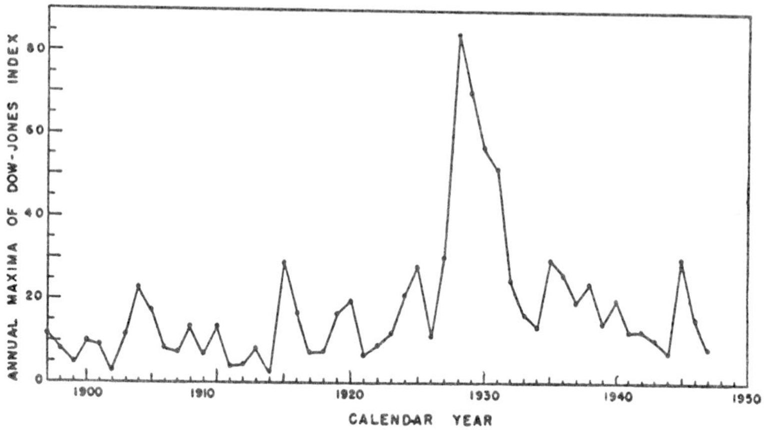

Fig. 11.9. Annual maxima, Dow-Jones Index, 1897-1947. From reference [9].

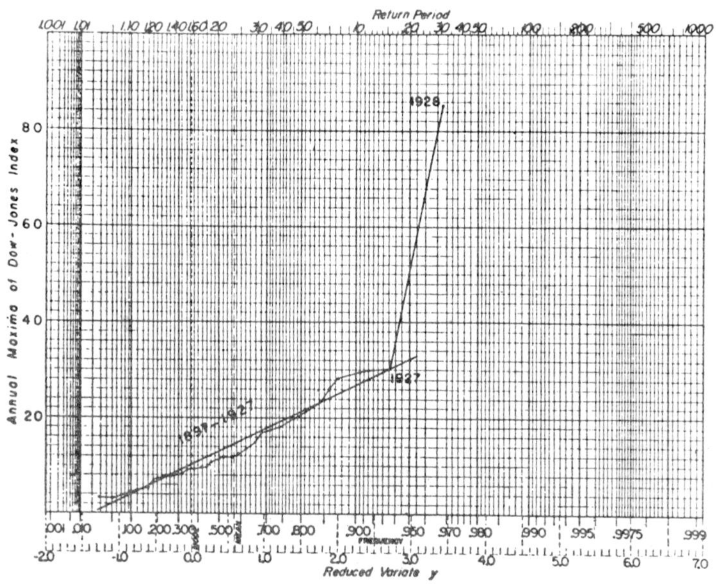

Fig. 11.10. Dow-Jones maxima plotted on Gumbel graph paper 1897-1928. From reference [9].

suggests that something changed in the basic mechanism in 1927, which is now known to be the case.

ACKNOWLEDGMENT

The research on the relationship between nonlinear differential and difference equations upon which part of the fourth section is based has been partially supported by the Fluid Dynamics Branch of the Office of Naval Research.

REFERENCES:

1. Beer, A. and Beer, P., Kepler, Vistas in Astronomy 18: 417 (1975).

2. Chandler, R. E., Herman, R. and Montroll, E. W., Oper. Res. 6· 165 (1958).

3. Edie, L. C., Foote, R. S., Herman, R. and Rothery, R. W., Traffic Eng. 33: 21 (1963).

4. Fisher, J and Pry, R., Practical Applications of Technological Forcasting in Industries, M. J. Cetron, editor, Wiley, 1971.

5. Gazis, D. C., Herman, R. and Potts, R. B., Oper. Res. 6: 499 (1959).

6. Gillmer, T. C., Modern Ship Design, Naval Inst. Press, Annapolis, Md., 1975.

7. Graunt, John, Natural and Political Observations Mentioned in a Following Index and Made Upon the Bills of Mortality, London, 1676; Reprint Edition, Johns Hopkins, 1936.

8. Greenburg, H., Oper. Res. 7: 79 (1959).

9. Gumbel, E. J., Statistical Theory of Extreme Values and Some Practical Applications, No. 33 NBS App. Math. Series, 1954.

10. Gumbel, E. J., Statistics of Extremes, Columbia Univ. Press, 1958.

11. Herman, R. and Montroll, E. W., Proc. Nat. Acad. Sci. 69: 3019 (1972).

12. Herman, R., Montroll, E. W., Potts, R. B. and Rothery, R. W., Oper. Res. 7: 86 (1959).

13. Montroll, E. W. and Badger, W. W., Introduction to Quantitative Aspects of Social Phenomena, Gordon and Breach, 1974.

14. Montroll, E. W., Proc. of First Symposium on Eng. Applic. of Random Function Theory and Prob. p. 231, John Wiley, 1963.

15. cf. Raymond Pearl, Studies in Human Biology (Baltimore 1924).

16. Prandtl, L. and Tietjens, O. G., Applied Hydro- and Aeromechanics, McGraw-Hill, 1934: Dover Reprint, 1957.

17. Smale, S. and Williams, R. F., J. of Math. Biol. 3: 1 (1977).

18. Smeed, R. J., J. of Transport Economics and Policy 2: 1 (1968).

19. Taylor, J. W. R., editor, The Lore of Flight, Crescent Books, New York, 1974.

20. Verhulst, P. F., Mem. Acad. Roy. Bruxelles 28: 1 (1844).

ELECTRICAL PHENOMENA IN DEVELOPING BIOLOGICAL SYSTEMS†

P. Ortoleva††
Department of Chemistry, Indiana University,
Bloomington, Indiana

ABSTRACT

Experiments indicate that a great variety of interesting developmental phenomena involve an important electrical component. Applied electrical fields are also known to strongly effect the course of development in some systems. These phenomena have been modeled mathematically in some cases. Here we present results on self-organization in single cell systems. Multicell systems are also briefly reviewed. These systems present interesting biological and mathematical problems.

I. EXPERIMENTAL EVIDENCE FOR THE ROLE OF ELECTRIC EFFECTS IN DEVELOPMENT

A. Overview

Over the past century a number of investigators have shown experimentally that electric effects are important in developing systems ([4], [9], [16]). The phenomena are of two types. Induced effects arise when an external field is imposed on the developing system, driving ionic currents in various directions dictated by the electrode or induction coil configuration. Autonomous phenomena are those for which the system generates its own electric field and concomitant ionic currents. The key in both phenomena is the degree to which the electric effects actually serve to control or otherwise effect development or whether they are a consequence of more fundamental, developmental processes. The experimental evidence indicates that both induced and autonomous effects mediate development in quite interesting and relevant ways. For example, it has been shown that applied fields can induce limb regeneration in frogs ([3], [28]) and young rodents ([14], [22]) wherein regeneration does not take place in the absence of an applied field. Autonomous fields have, for example, been found in Fucus eggs [9], barley root hairs [9], Acetabularia [18] and fingertip regeneration in children ([5], [8]). A number of other examples are discussed in the reviews cited above. It is clear from the breadth of system types--ranging from single cells to whole organisms--that a number of rather different phenomena and mechanisms are involved here.

---

† Work supported in part by a grant from the National Science Foundation.
††A. P. Sloan Fellow, 1980-1982.

B. Early *Fucus* Egg Development--A Model Self-Organizing System

The case for the importance of self-organization of a cellular electrophysiological system is made most easily with a discussion of the early development of the egg of the seaweed *Fucus*. After fertilization, this egg is spherical in shape with nucleus at the origin and a spherically symmetric distribution of all other cell features. Yet after the first division it is left with two dramatically different progeny--the rhizoid that eventually develop into the roots and the thallus that becomes the leaf-stem system. How is this system able to create this specialization?

It has long been recognized that this development of an asymmetric division had a strong electrical character [15]. More recently it was shown that asymmetry sets in at an early stage in the form of a pattern of voltage and concomitant currents as seen in Figure 12.1 (see Jaffe, [9], for a review). This development of electrophysiological pattern formation was shown to be randomly oriented in dark and gradient free culture medium but could be oriented by applied electrical fields (known from the work of Lund [15]) and the presence of composition gradients, proximity to other eggs or centrifugation. The centrifugation experiments [31] showed that cytoplasmic streaming did not play a role in axis formation and that the direction of axis formation after centrifugal stratification of the egg could be reversed by changing the composition of the culture medium, notably pH. Important findings that eliminated the role of microtubules or even appreciable protein synthesis were made by Quatrano [25].

These factors were conjectured to be strong evidence for the fact that early *Fucus* development was an example of a Turing [30] like mechanism ([19], [20]). Some interesting suggestions for mechanisms that could lead to positive feedback capable of sustaining polarity in *Fucus* were made by Jaffe and coworkers (see Jaffe, [9], for a review) and were put into a unified self-consistent model by Larter and Ortoleva ([10], [11], [12]) and Larter, Strickholm and Ortoleva [13].

The initial symmetry of the *Fucus* egg is very important since it enables us to make a strong case for a self-organization phenomenon. It should not be used as an argument to say that *Fucus* is then so atypical as to be uninteresting. Indeed, if self-organization can take place in *Fucus* eggs then it can even much more easily happen in cells of some degree of initial asymmetry such that the initial asymmetry is greatly amplified by the pattern forming mechanism as we will discuss in Section V below.

II. THEORIES OF DEVELOPMENTAL BIOELECTRICITY

That electrical effects should be important in development is actually quite obvious. One essential element of development is the distribution of various molecules

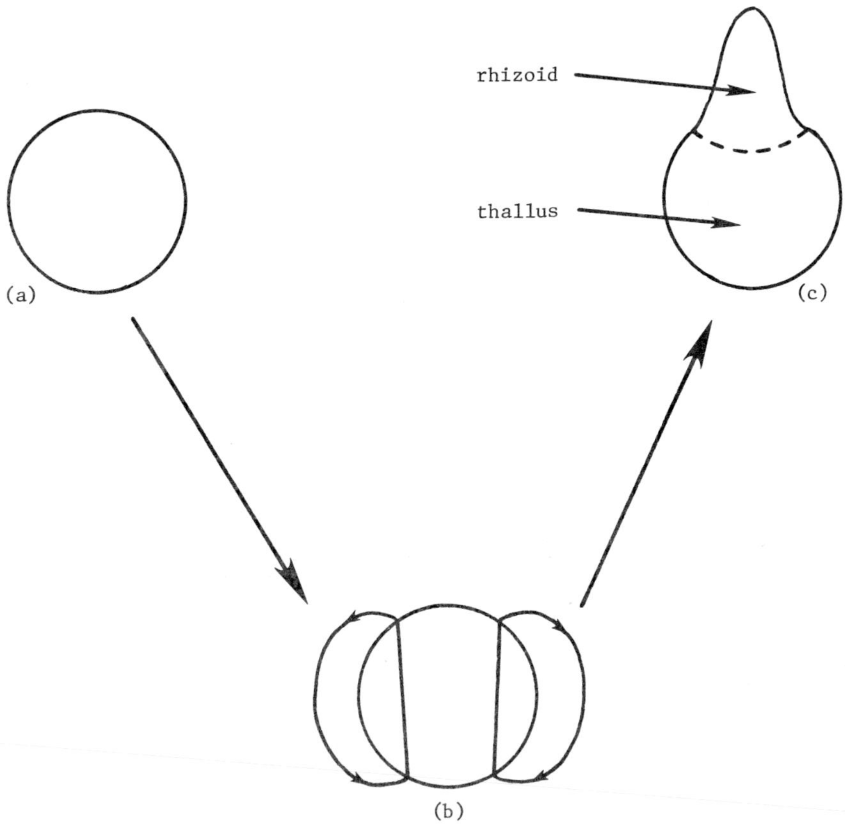

Fig. 12.1. Polarization of Fucus egg. The initially spherically symmetric egg with symmetrically distributed membrane potential spontaneously develops a transcellular flux of ions within an hour after fertilization. Arrows indicate the flux of Ca^{2+}. The egg will eventually elongate in the direction of the current and undergo asymmetric division perpendicular to the axis defined by the transcellular field. It is believed that the spontaneously generated electrophysiological asymmetry is the fundamental event that sets the stage for the later asymmetric division into the rhizoid and thallus cells that are the presumptive root and leaf-stem systems respectively.

from one region of the system to an appropriate target. Since many, if not most, biologically relevant chemical species are ionic (from Na^+, K^+, and Cl^- to proteins) it is clear that the imposition of electric fields can aid or interfere with the "normal" targeting of ions.

In the case of autonomous phenomena the situation is more subtle. Biomembranes have the very interesting property that they can pump ions and nonionic species across themselves against an adverse chemical potential difference. Thus they can act as ion current sources leading to net electric currents and associated ohmic voltage gradients. Also there is the strong tendency towards charge neutrality leading to voltage gradients that coordinate the ionic motions so that charge neutrality is maintained. To this we now add the consideration of Turing's biomorphogenetic hypothesis [30]. Turing proposed that the instability of the symmetric state of some reaction-diffusion system to pattern forming composition perturbations could account for the ability of certain biosystems to spontaneously self-organize as they are typically observed to do during development. Perhaps the most amusing study along these lines is that by Murray [17] wherein the development of animal coat markings (such as that of a zebra) can be explained qualitatively by a Turing-type theory. Recent work, to be outlined here, has shown that some autonomous bioelectric effects in development can be explained by adding the dimension of electrical effects to the Turing theory. The great variety of types of systems showing interesting developmental bioelectric effects makes it clear that there is no one underlying mechanism or theory to explain the observed phenomena. Since theory in this area is in a fairly early stage, we shall restrict our considerations to unicellular phenomena where most detailed calculations have been made. Some results on multicellular systems are discussed in Ortoleva [21].

III. EQUATIONS OF CELLULAR DEVELOPMENTAL BIOELECTRICITY

A mathematical description of a self-organization phenomenon in a cell requires an accounting of the distribution of chemical composition and voltage at all points in the system. A complete theory of this type does not require the need to know, say, the voltage to calculate the concentrations; all quantities must be calculated self-consistently. To illustrate how such a theory can explain the manner in which electric fields can lead to autonomous and induced developmental phenomena we consider a simple intuitive model involving membrane transport feedback leading to self-organization ([21], [27]). The model is based on some ideas of Jaffe and co-workers (see Jaffe [9] and Andersson [1]) and given a self-consistent mathematical formulation by Larter and Ortoleva ([10], [11], [12]) and Larter, Strickholm and Ortoleva [13].

A. Physical Arguments for a Mobile Membrane Transport Mediator (MMTM) Model

It has been shown experimentally ([23], [24]) that large molecules may move along a membrane under the action of electric fields believed to act in *Fucus* ([9], [12]). Thus if ion pumps or leaks could move along the membrane in this manner, a simple feedback loop leading to self-organization can occur in the form of patterns of concentration of the pumps and leaks on the outer membrane.

To be more explicit, suppose a Ca^{2+} extrusion pump may move along the outer cell membrane. Assume this pump is positively charged, and the part sticking out of the membrane into the cell interior is positive. Then if a few extra Ca^{2+} pumps are in some region of the membrane, then Ca^{2+} is more rapidly pumped out of the cell leaving that region negative (near the membrane in the cell interior) relative to other regions in the cell along the cell membrane. But since the Ca^{2+} pump is positively charged on the interior side, more Ca^{2+} pumps come into the original region and there is a runaway process that accentuates the nonuniformity. Thus the state of uniform Ca^{2+} pump distribution can become unstable to pattern forming perturbations of Ca^{2+} pump surface concentration.

To formulate such a picture mathematically we first assume the existence of molecules or complexes that affect the flux of ions across the membrane. They are distributed with concentrations $\underline{\Gamma} = \{\Gamma_1, \Gamma_2, \ldots, \Gamma_B\}$ (moles/area) on the cell membrane. Letting $\underline{\Gamma}^h$ be the uniform distribution value of $\underline{\Gamma}$, we assume that the voltage δV^i in the cell interior evaluated at the cell membrane induced by an inhomogeneous distribution $\delta\underline{\Gamma} = \underline{\Gamma} - \underline{\Gamma}^h$, can be written

$$\delta V^i = \sum_{\alpha=1}^{B} \kappa^i_\alpha \delta\Gamma_\alpha, \tag{12.1}$$

i.e., $\delta\underline{\Gamma}$ induces a voltage proportional to itself. The κ^i are expected to reflect the fact that δV^i at any point along the membrane should depend on the distribution of $\delta\underline{\Gamma}$ over the entire surface. Hence κ^i is an integral operator. Letting $\underline{\xi}$ be a set of two curvilinear coordinates specifying points on the cell surface we have

$$[\kappa^i_\alpha \Psi](\underline{\xi}) = \int d^2\omega' K^i(\underline{\xi}, \underline{\xi}')\Psi(\underline{\xi}') \tag{12.2}$$

for any function $\Psi(\underline{\xi}')$ that varies over the cell surface. Here $\int d^2\omega$ indicates an integration over the cell surface. A similar result for the voltage deviation δV^e along the outer surface of the cell membrane can be postulated introducing operators κ^e_α. The explicit calculation of the functions $K^{i,e}_\alpha(\underline{\xi}, \underline{\xi}')$ from cell properties is discussed in Section IIIB below. Physically $\kappa^{i,e}_\alpha$ indicates the strength of the feedback for $\underline{\Gamma}$ self-organization as we shall see.

To complete the theory we invoke an equation of mass conservation for $\underline{\Gamma}$. For simplicity we assume that $\underline{\Gamma}$ is not affected by chemical reactions. For example, pumps are assumed driven by a well buffered ATP concentration. With this we get

$$\frac{\partial \underline{\Gamma}}{\partial t} = -\vec{\nabla}_s \cdot \vec{\underline{j}} \tag{12.3}$$

where $\vec{\underline{j}}$ are the $\underline{\Gamma}$-fluxes along the membrane; $\vec{\nabla}_s$ is a gradient operator along the membrane surface. If the Γ_α are not too large then we expect

$$\vec{j}_\alpha = -d_\alpha \vec{\nabla}_s \Gamma_\alpha - m_\alpha^i \Gamma_\alpha \vec{\nabla}_s \delta V^i - m_\alpha^e \Gamma_\alpha \vec{\nabla}_s \delta V^e \tag{12.4}$$

where $-\vec{\nabla}_s V^i$ and $-\vec{\nabla}_s V^e$ are the electric fields along the membrane at the interface of the membrane and regions i and e respectively (see Figure 12.2); d_α is the diffusion coefficient of α along the membrane. The mobility factors m_α^i, m_α^e are different

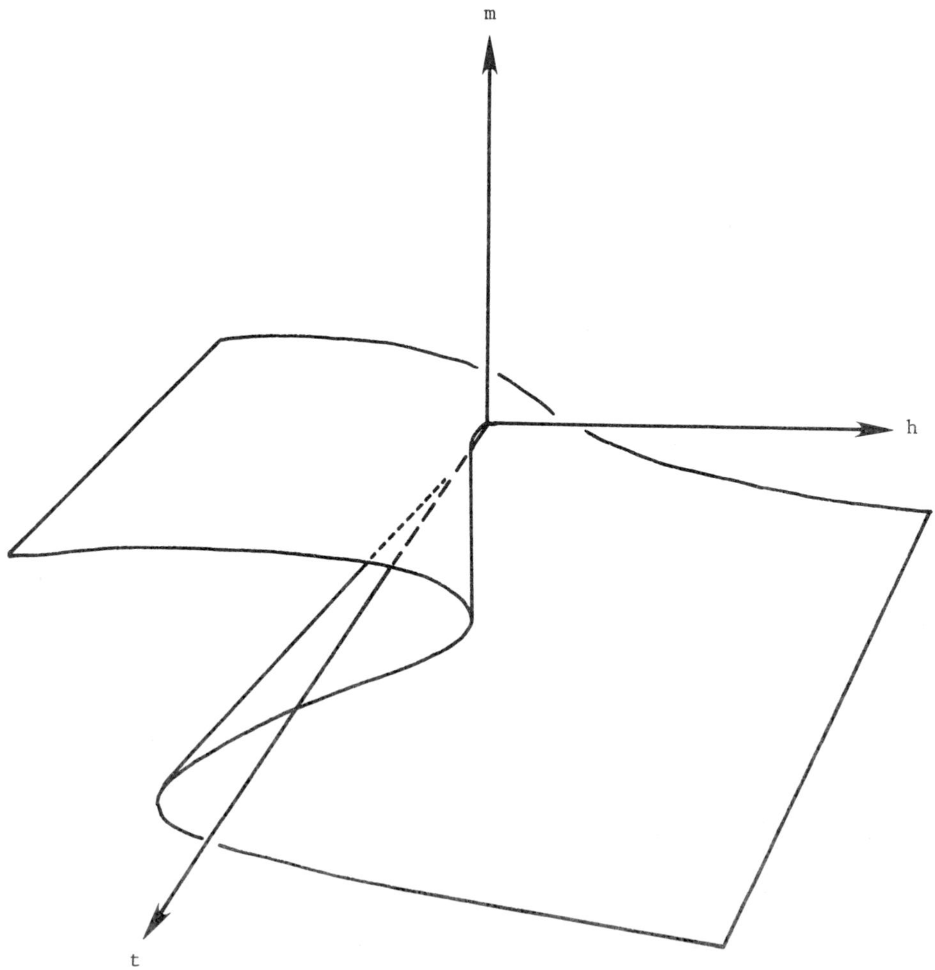

Fig. 12.2. Nonequilibrium phase diagram for magnitude m of transcellular current as a function of the time t after which asymmetry sets in under the influence of an applied electric field of magnitude h. The cusp surface in t-h-m space is essentially the same as the phase diagram for a ferromagnet where m, t and h are the magnetization, deviation of temperature from the Curie temperature and the applied magnetic field respectively.

to reflect the fact that the ends of α sticking into i or e may have different valences z_α^i, z_α^e and hence the mobilities must be broken up into a bare part m_α (> 0) and the valence factor ($m_\alpha^i = z_\alpha^i m_\alpha$, $m_\alpha^e = z_\alpha^e m_\alpha$). Einstein's relation would suggest that d_α and m_α are related by $d_\alpha = RTm_\alpha/\mathcal{F}$ where R is the gas constant, T is the absolute temperature and $\mathcal{F}$ is Faraday's constant.

The central result of these handwaving arguments is to obtain a closed equation for $\underline{\Gamma}$. This is accomplished by introducing an operator $\lambda\kappa_\alpha$ such that

$$\lambda\kappa_\alpha \equiv m_\alpha^i \kappa_\alpha^i + m_\alpha^e \kappa_\alpha^e \tag{12.5}$$

where the parameter λ is introduced to be a formal measure of the feedback strength. With this (12.1) yields

$$\frac{\partial\Gamma_\alpha}{\partial t} = \vec{\nabla}_s \cdot \left[d_\alpha \vec{\nabla}_s \Gamma_\alpha + \lambda\Gamma_\alpha \vec{\nabla}_s \sum_{\beta=1}^{B} \kappa_\beta \Gamma_\beta \right] \tag{12.6}$$

where we have used the fact that a uniform distribution $\underline{\Gamma}^h$ leads to no net field along the membrane, i.e., $\vec{\nabla}_s \kappa_\alpha \Gamma_\alpha^h = \vec{0}$. (12.6) is a closed equation for $\underline{\Gamma}$ and hence presents itself as a self-consistent model--i.e., we do not need to postulate an electric field to calculate an inhomogeneous distribution of $\underline{\Gamma}$; the theory, via (12.1-12.6) calculates all relevant quantities.

B. Derivation of the MMTM Model

A more complete derivation of the above model both helps to clarify some of the assumptions underlying it and yields an explicit relation between κ and the cell rate and transport processes. This derivation starts from reaction-electrodiffusion equations for the cell interior (i) and exterior (e) regions (assumed to be continuous media), charge neutrality conditions and boundary conditions relating fluxes in i and e normal to the outer membrane to the membrane localized reaction and transmembrane flux rates ([11], [27]). To obtain the model of the previous section we must assume

(1) The motion of $\underline{\Gamma}$ along the membrane is the rate limiting step in the developmental process so that the dynamics in i and e are at quasi-steady state consistent with a given $\underline{\Gamma}$ distribution.

(2) The gradients in i and e are sufficiently small that deviations from the unpatterned state may be linearly related to the deviation $\delta\underline{\Gamma}$ of $\underline{\Gamma}$ from the unpatterned $\underline{\Gamma}$ distribution $\underline{\Gamma}^h$.

A direct implication of this derivation is that unless at least one of the species whose transmembrane flux is effected by $\underline{\Gamma}$ must be only weakly mobile in the cell interior. The prime candidate here is Ca^{2+} which tends to bind strongly to sites fixed to membraneous and other immobile intracellular structures.

The formation of pattern has been studied without making assumptions (1) and (2) above ([12], [13]), and comparison with the simplified theory indicates that the latter retains the qualitative and much of the quantitative picture to good approximation. (More mathematical details are given below.)

C. Other Cell Electrophysiological Self-Organization Models

The MMTM model is but one of a number of models that have been invoked to describe self-organization in cells. Other models are as follows and are reviewed in Ortoleva [21].

1. Classic Turing Models

In Turing's original paper, pattern formation resulted from an interplay between bulk reaction and diffusion. This is in contrast to the MMTM model wherein the important feedback for the self-organization is located at the system boundary--the cell membrane. Theories of the formation of cellular polarity (and other patterns) more closely in the spirit of Turing's original hypothesis are found in Ortoleva and Ross ([19], [20]) and Billing and Hunding [2].

2. The Release Membrane Incorporation Model

At various stages of cellular development certain molecules are released or rapidly but transiently produced. If these molecules can be immobilized at the outer membrane, are charged and effect the transport of charged species across the cell membrane, then a self-organizing feedback loop can clearly be set up [21].

3. Membrane Transport Nonlinearity

If the membrane transport of one ion is mediated by itself or another, then a pattern formation tendency is possible. Such a model involving Ca^{2+} has been demonstrated to lead to patterning for a Ca^{2+} cell model [10].

In the above list of scenarios for cellular pattern formation, electric fields played a key (MMTM and #2) or a secondary (#1, 3) role. Of course the electrical character of the species involved (#1, 3) can profoundly effect specific conditions under which spontaneous patterning is possible.

IV. STABILITY AND PATTERN FORMATION

The self-consistent set of equations (12.6) provides a formalism for investigating questions of self-organization of charged membrane bound species. In this section we examine how these effects are manifest in this model. We specialize to the case of one species. Here we stress the physics of the problem, details of the calculation being found in Smiley and Ortoleva [27].

A. Steady States

The steady states of the Γ distribution are those for which Γ is independent of time. Since Γ is, by assumption, not affected by any reactions this implies that $\vec{\nabla}_s \cdot \vec{j} = 0$ which for the one species case is satisfied if $\vec{j} = \vec{0}$. Hence letting λ/d be denoted simply λ, (12.6) implies

$$\ell n\Gamma + \lambda\kappa\Gamma = H \tag{12.7}$$

where H is a constant that must be fixed by invoking the condition

$$\int d^2\omega\Gamma = \int d^2\omega\Gamma^h = A \tag{12.8}$$

which states that the total amount of Γ is conserved so that this total must be that in the unpatterned (uniform) state. The last part of (12.8) indicates that we can always choose units for Γ such that $\Gamma^h = 1$; A is then seen to be the cell surface area. The constant H may be found explicitly by exponentiating both sides of (12.7) to get $\Gamma = \exp[H - \lambda\kappa\Gamma]$ and invoking (12.8). This yields

$$\ell n\Gamma + \lambda\kappa\Gamma = \ell n\left[\frac{A}{\int d^2\omega e^{-\lambda\kappa\Gamma}}\right]. \tag{12.9}$$

Noting that $\kappa \cdot 1$ is a constant, i.e., we assume no transcellular fields when $\Gamma = \Gamma^h \equiv 1$, we see that Γ^h satisfies this steady state equation.

B. Instability of the Unpatterned State

The tendency towards spontaneous pattern formation manifests itself in the answer to the stability question: Is the unpatterned state unstable to small deviations of a given pattern symmetry? We proceed by letting $\delta\Gamma$ be the deviation from the unpatterned state, i.e., $\delta\Gamma = \Gamma - 1$, and then retain only linear terms in $\delta\Gamma$ in the equation

$$\frac{\partial\Gamma}{\partial t} = \nabla_s^2\Gamma + \lambda\vec{\nabla}_s \cdot \Gamma\vec{\nabla}_s\kappa\Gamma \tag{12.10}$$

to obtain

$$\frac{\partial\delta\Gamma}{\partial t} = \nabla_s^2[1 + \lambda\kappa]\delta\Gamma. \tag{12.11}$$

Since κ is time independent (see (12.2)), $\delta\Gamma$ may be written in separable form $\delta\Gamma = e^{\zeta t}\hat{\Gamma}(\underline{\xi})$ where

$$\{\nabla_s^2[1 + \lambda\kappa] - \zeta\}\hat{\Gamma} = 0. \tag{12.12}$$

Thus the system becomes unstable to patterns when ζ passes from negative to positive. To organize the analysis rather neatly, we shall study its behavior as the feedback strength λ changes. Thus the conditions for marginal stability, $\zeta(\lambda_c) = 0$

yields values λ_c when

$$[1 + \lambda_c \kappa]\hat{\Gamma} = 0 \tag{12.13}$$

has nontrivial solutions. Thus if k_α are the eigenvalues of κ, i.e., there are a set of Ψ_α's such that

$$\kappa\Psi_\alpha = k_\alpha \Psi_\alpha, \tag{12.14}$$

then we see

$$\lambda_c(\alpha) = -1/k_\alpha, \tag{12.15}$$

i.e., the system is marginally stable to the formation of patterns $\Psi_\alpha(\underline{\xi})$ when $\lambda = \lambda_c(\alpha)$. Since, according to Section IIIB, κ can be obtained in terms of the cell transport and reaction rate coefficients and culture medium conditions, this gives an explicit calculation of the conditions for pattern formation in terms of the biological parameters (in the particular combination k_α).

C. Bifurcation of Polarity

An important idealization of the early Fucus egg development is that it reaches a steady polarized state. Such states manifest themselves in the present model as nonuniform solutions of (12.9).

For a spherical cell one can show that the kernel K associated with κ takes the form

$$K(\theta\phi, \theta'\phi') = \sum_{\ell=0}^{\infty} \sum_{m=-\ell}^{\ell} k_\ell Y_{\ell m}(\theta\phi) Y^*_{\ell m}(\theta'\phi') \tag{12.16}$$

where $\theta\phi$ are spherical angles, the $Y_{\ell m}(\theta\phi)$ are the spherical harmonics and the k_ℓ can be calculated from cell parameters. It is clear that

$$\kappa Y_{\ell m} = k_\ell Y_{\ell m} \tag{12.17}$$

and hence ℓ-patterns are marginally stable at the critical values $\lambda_c(\ell) = -1/k_\ell$. Furthermore the associated patterns of marginally stable perturbations at $\lambda = \lambda_c(\ell)$ are the $2\ell + 1$ modes $Y_{\ell m}$, $m = -\ell, -\ell + 1, \ldots, 0, \ldots, \ell - 1, \ell$. In fact, it is easy to show that

$$\zeta_\ell(\lambda) = -\ell(\ell + 1)[1 + \lambda k_\ell]. \tag{12.18}$$

Hence the system is unstable to ℓ-perturbations when $\lambda < -1/k_\ell$, $k_\ell > 0$, $\lambda > -1/k_\ell$, $k_\ell < 0$.

The polarized state with current loops as in Fig. 12.1(b) is an $\ell = 1$ pattern; it should be the first to arise as $|\lambda|$ is increased since it corresponds to the longest wavelength, minimizing the effect of diffusion to stabilize patterned perturbations.

To calculate the polar state perturbatively for λ near $\lambda_c(1)$ we may invoke standard bifurcation theory techniques ([6], [26], [29]). To perform the calculation we must introduce an amplitude parameter ε which is small for $\lambda \sim \lambda_c(1)$. Thus we assume the existence of expansions in the form

$$\gamma \equiv \Gamma - 1 = \sum_{n=1}^{\infty} \gamma_n \varepsilon^n \tag{12.19}$$

$$\lambda - \lambda_c(1) = \sum_{n=1}^{\infty} \lambda_n \varepsilon^n . \tag{12.20}$$

The reason why we do not expand directly in powers of $\lambda - \lambda_c(1)$ is that often, and in fact for the polar state of interest here, the amplitude of the pattern is of order $[\lambda - \lambda_c(1)]^{1/2}$ or some other fractional power of $\lambda - \lambda_c$; (12.20) is the aspect of the present procedure that allows us to build this possibility into the theory.

The results of this bifurcation theoretic calculation are that $\lambda_1 = 0$, λ_2 can be chosen to be unity (for convenience) and

$$\gamma_1 = \sum_{m=-1}^{1} g_m Y_{1m} \tag{12.21}$$

$$\begin{aligned} g_0(g_0^{\,2} + 2g_1 g_{-1} - f\lambda_2) &= 0 \\ g_1(g_0^2 + 2g_1 g_{-1} - f\lambda_2) &= 0 \\ g_{-1}(g_0^2 + 2g_1 g_{-1} - f\lambda_2) &= 0 \end{aligned} \tag{12.22}$$

where f can be related to k_0, k_1 and k_2. Using the definitions of the spherical harmonics it is easy to show that

$$\gamma \underset{\lambda-\lambda_c(1)}{\sim} \sqrt{\frac{3f(\lambda - \lambda_c(1))}{4\pi}} \cos \chi(\hat{n}) \tag{12.23}$$

where $\hat{n}$ is the unit vector along the z-axis of an arbitrarily oriented coordinate system and χ is the angle between $\hat{n}$ and the ray to the point of interest on the sphere. The existence of a continuous infinity of solutions is an obvious consequence of the spherical symmetry of the *Fucus* egg and explains why the axis is randomly oriented when grown in a gradient free medium.

D. Pattern Stability and Secondary Bifurcations

It is sometimes observed that *Fucus* eggs develop a double rhizoid. At first one might think that this corresponds simply to the onset of an $\ell = 2$ pattern. However, in their investigation of a Turing model of pattern formation in a sphere (neglecting electrical effects), Hunding and Billing [7] showed that an ($\ell = 2$)-like pattern could arise from the instability of an $\ell = 1$ pattern. We are investigating dirhizoid development in the present MMTM model.

V. CELL ASYMMETRIES, APPLIED FIELDS AND IMPERFECT BIFURCATION

The spherical symmetry of the Fucus egg is more of an anomaly than a typical situation. Also development often takes place in an inhomogeneous environment. Thus we investigate these imposed symmetry breaking effects in systems with the potentiality for self-organization. Also, these calculations are of interest as a guide to unambiguously interpret the results of applied field experiments on developing cells as a probe of differentiation mechanisms. We now demonstrate according to the MMTM model.

A. Modification of the MMTM Theory

Generalizations of the MMTM model equations to include applied fields and inherent cell asymmetry can be done rigorously ([12], [27]). Here we keep to an intuitive development. The key step in setting up the MMTM theory of Section IIIA was the assumption that electric fields were a linear response to the asymmetry of the distribution of $\underline{\Gamma}$ over the cell surface, see (12.1). However, in the presence of an applied electric field, here assumed to be constant of magnitude E in the z direction at infinite distance from the cell, there is an electric field even if $\underline{\Gamma}$ is distributed uniformally. Thus for a spherically shaped cell we must add a term $-E\cos\theta$ to (12.1). Similarly, if there are specialized regions on the cell surface or interior which are asymetrically distributed, such as membrane pores or special compartments as in amphibian eggs, then we might expect a second additional term F in (12.1) where F reflects the pattern of the asymmetry and can in principle depend on Γ as well as on the position $\underline{\xi}$ on the membrane. With these considerations we expect that (12.7) for a spherical shaped cell is generalized to an equation of the form, for one MMTM species,

$$\ell n\Gamma + \lambda\kappa\Gamma + bF + cE\cos\theta = H \qquad (12.24)$$

for the steady distributions, where b and c are easily calculated constants. Again H is determined as a functional of Γ by invoking the constancy of the total amount of the MMTM species, i.e., (12.8). Note that $\Gamma = 1$ is no longer a steady state of the inherently asymmetric problem. Now we address the question of self-organization in the presence of this asymmetry and how the latter interferes with or augments this ordering process.

B. Imperfect Bifurcation

The effect of applied and inherent asymmetry to a cell of spherical shape may easily be calculated if the asymmetry is small. In particular we consider the effect on the $\ell = 1$ pattern considered in Section IVC. We assume E and b are small such that

$$E = E_3\varepsilon^3, \quad b = b_3\varepsilon^3. \tag{12.25}$$

Using a bifurcation analysis as in Section IVC, it was found for $\lambda \sim \lambda_c(1)$ that the emergent pattern is given by (12.21) but that the amplitudes g_m are given by

$$g_0(g_0^2 + 2g_1g_{-1} - f\lambda_2) = \tilde{b}_0F_{10} + \tilde{c}E_3 \tag{12.26}$$

$$g_1(g_0^2 + 2g_1g_{-1} - f\lambda_2) = \tilde{b}_1F_{11} \tag{12.27}$$

$$g_{-1}(g_0^2 + 2g_1g_{-1} - f\lambda_2) = \tilde{b}_{-1}F_{1-1} \tag{12.28}$$

where $\{\tilde{b}_m\}$ and $\tilde{c}$ are related to the constants of the problem and $F_{\ell m}$ is defined as

$$F_{\ell m} = \int_0^{2\pi} d\phi \int_0^{\pi} \sin\theta d\theta Y^*_{\ell m}(\theta, \phi)F(\Gamma = 1; \theta, \phi) \tag{12.29}$$

where $\Gamma = 1$ is, in the present units, the uniform MMTM distribution over the sphere.

1. Symmetric Cell in an External Field

For the symmetric cell $F = 0$. It then is easy to show that the only solution to (12.26-12.28) is $g_1 = g_{-1} = 0$ and g_0 satisfies

$$g_0(g_0^2 - f\lambda_2) = \tilde{c}E_3. \tag{12.30}$$

To understand this result we multiply through by ε^3 and get

$$m[m^2 - f(\lambda - \lambda_c(1))] = \tilde{c}E \tag{12.31}$$

where $m = \varepsilon g_0$ and hence $\gamma \sim mY_{10} = \varepsilon g_0\sqrt{\frac{3}{4\pi}}\cos\theta$. Letting

$$\begin{aligned} t &\equiv -f(\lambda - \lambda_c(1)) \\ h &= -\tilde{c}E \end{aligned} \tag{12.32}$$

the solutions to (12.31) form a cusp in (m, t, h) space (see Figure 12.2). This topology is the same as that for a ferromagnetic transition where m, t, h represent the magnetization, deviation of the temperature from the Curie temperature and the applied magnetic field respectively.

2. Interplay of Internal Asymmetry and External Gradients

Fucus eggs stratify their internal components under centrifugation [31]. The centrifugal direction then determines the axis of *Fucus* polarity after subsequent development. However, if the cell is also subjected to external gradients, it is the interplay of these factors that determines the direction of axis formation. This phenomenon is manifest in the interaction of the $F_{\ell m}$ and E_3 terms in (12.26-12.28). This and the applied field effects are discussed further in Smiley and Ortoleva [27].

C. Hyperpolarizability

A ferromagnet exhibits an anomalous response to a change of temperature or magnetic field as can be seen from the divergence of the slope of the surface m(t, h) in Figure 12.2 near t = 0, h = 0, the cusp point. Thus by analogy the same phenomenon is seen to occur here. A cell subject to an external field E responds strongly. This "hyperpolarizability" phenomenon [21] is similar to the large response to external influences found in other nonequilibrium phase transitions ([19], [20]). Note that the response comes in a very well defined way--i.e., like the field free pattern except for its orientation. This allows a biosystem to be sensitive to its environment but with a response that is well controlled by the cell's internal kinetics.

D. Lability

The direction of polar axis fixation of Fucus can be rotated in early development by changing the direction of an external gradient. This allows the system to adapt to changes in its environment. In the present theory this results from the marginal stability of the $\ell = 1$ pattern to axis rotation-type perturbation.

VI. OTHER ELECTRICAL PHENOMENA IN DEVELOPING SYSTEMS

Experiments show a great variety of autonomous current and applied field phenomena are of great significance in developing systems ([4], [9], [16]). Mathematical modeling of some of these phenomena are reviewed by Ortoleva [21]. Multicellular systems, wave propagation and profound effects of system geometry are discussed there.

It is clear that this area of research is at the juncture of biology, physical chemistry and mathematics. It is important because it addresses both fundamental questions regarding the nature of life and the need for promoting regeneration in humans. In the latter case, electrical promotion of bone healing is already an accepted medical practice yielding results in cases where bones have not healed normally [4].

ACKNOWLEDGMENTS

I thank my coworkers J. Chadam, M. El Badwei, D. Feinn, R. Larter, S. Schmidt, M. Smiley and A. Stirckholm, without whose collaboration this work would not have been possible.

REFERENCES:

1. Andersson, G., Self-electrophoresis of membrane molecules and chromosomes: Unifying hypothesis of cell cycle controls, J. Theor. Biol. 77: 1 (1979).

2. Billing, G. and Hunding, A., Bifurcation analysis of nonlinear reaction-diffusion systems: Dissipative structures in a sphere, J. Chem. Phys. 69: 3603 (1978).

3. Bodemer, C. W., Evocation of regrowth phenomena in anuran limbs by electrical stimulation of the nerve supply, Anat. Rec. 148: 441 (1964).

4. Brighton, C. T., Black, J. and Pollack, S. R., editors, Electrical Properties of Bone and Cartilage: Experimental Effects and Clinical Applications, Grune and Stratton, N.Y., 1979.

5. Douglas, B. S., Conservative management of guillotine amputation of the finger in children, Austr. Paediatr. J. 8: 86 (1972).

6. Gurel, O. and Rossler, O., editors, Proc. of the Conf. on Bifurcation Theory and its Applications, Ann. N.Y. Acad. Sci., 316: (1979).

7. Hunding, A. and Billing, G., Secondary bifurcations in spherical reaction-diffusion systems, Chem. Phys. 45: 359 (1980).

8. Illingworth, C. M., Trapped fingers and amputated finger tips in children, J. Pediatr. Surg. 9: 853 (1974).

9. Jaffe, L. F., Control of development by ionic currents, in Membrane Transduction Mechanisms, Cone, R. A. and Dowling, J. E., editors, Raven Press, N.Y., Soc. Gen. Physiol. Ser. 33: 199 (1979).

10. Larter, R. and Ortoleva, P., A theoretical basis for self-electrophoresis, J. Theor. Biol., in press.

11. Larter, R. and Ortoleva, P., A study of instability to electrical symmetry-breaking in unicellular systems (submitted for publication).

12. Larter, R. and Ortoleva, P., A bifurcation theory for the study of electrical effects in biological systems and the effect of external fields on the bifurcation diagram (submitted for publication).

13. Larter, R., Strickholm, A. and Ortoleva, P., A detailed model of the electrical symmetry-breaking transition in early Fucus egg development (submitted for publication).

14. Libbin, R. M., Person, P., Papierman, S., Shah, P., Nevid, D. and Grob, H., Partial regeneration of the above-elbow amputated rate forelimb: II. Electrical and mechanical facilitation, J. Morph. 159: 427 (1979).

15. Lund, E. J., Electrical control of organic polarity in the egg of Fucus, Botan. Gazette 76: 288 (1923).

16. Lund, E. J., Bioelectric Fields and Growth, U. of Texas Press, Austin, 1947.

17. Murray, J., A Turing theory to animal coat patterns, J. Theoret. Biol., to appear; 1980.

18. Novak, B. and Bentrup, F. W., An electrophysiological study of regeneration in Acetabulana mediterranea, Planta 108: 227 (1972).

19. Ortoleva, P. and Ross, J., A chemical instability mechanism for asymmetric cell differentiation, J. Biophys. Chem. 1: 87 (1973).

20. Ortoleva, P. and Ross, J., A theory of asymmetric cell division, Dev. Biol. 34: F19 (1973).

21. Ortoleva, P., Developmental Bioelectricity, Proceedings of the Conference on Nonionizing Radiation, ACS Symposium Series, Houston, Texas, March 1980 Meeting.

22. Person, P., Libbin, R. M., Shah, D. and Papierman, S., Partial regeneration of the above-elbow amputated forelimb: I. Innate responses, J. Morph. 159: 427 (1979).

23. Poo, M.-m., Poo, W-J. H. and Lam, J. W., Lateral electrophoresis and diffusion of concanovalian A receptors in the membrane of embryonic muscle cell, J. Cell. Biol. 76: 483 (1978).

24. Poo, M.-m. and Robinson, K. R., Electrophoresis of concanovalian A receptors along embryonic muscle cell membrane, Nature 265: 602 (1977).

25. Quatrano, R. S., Separation of processes associated with differentiation of two-celled Fucus embryos, Dev. Biol. 30: 209 (1973).

26. Sattinger, D., Topics in Stability and Bifurcation Theory, Springer-Verlag, New York, 1973.

27. Smiley, M. and Ortoleva, P., Bifurcation of current patterns in developing cells (in preparation).

28. Smith, S. D., Induction of partial limb regeneration in Rama pipiens by galvanic stimulation, Anat. Rec. 158: 89 (1974).

29. Statgold, I., Branching of solutions of nonlinear equations, SIAM Rev. 13: 289 (1971).

30. Turing, A. M., The chemical basis of morphogenesis, Phil. Trans. Roy. Soc., London, Ser. B 237: 37 (1952).

31. Whitaker, D. M., The effects of centrifugation on developing Fucoid eggs, J. Cell Comp. Physiol. 15: 173 (1940).

NONLINEAR DIFFUSION EQUATION IN BIOLOGY

Ei Teramoto and Nanako Shigesada
Department of Biophysics, Kyoto University,
Kyoto, Japan

INTRODUCTION

Mathematical formulation of Brownian motion originated by Einstein [2] and Smoluchowski [11] was completely confirmed by the observations of a number of colloidal particles in a well-defined visual area of the ultramicroscope (Svedverg [12] and Westgren [14] (1916)), and now the diffusion equation based on the theory of Brownian motion is used as the fundamental formula for the quantitative description of stochastic phenomena in a wide range of scientific fields. In the case of Brownian motion caused by the random forces of an external field, the formulation could be deduced from first principles: laws of classical mechanics, thermodynamics and probability theory.

On the contrary, the formulation of dispersive motion of living organisms can by no means be deduced from the first principles, unless we consider such the exceptional cases as simple passive dispersions of spores and phytoplanktons. Really, the active dispersive motions of animals also show the variety of behavioral patterns depending upon the species characteristics: migration of migratory birds, schooling of fish, swarming of insects and so on. Thus, in the present discussion, we must direct our attention to rather simple dispersive motions of small organisms. Even in those cases, we have no guiding laws based on which the formulation can be logically deduced, so all we can do is start from the analyses of the statistical data of some experimental observations.

For the present purpose, the dexterous experiments of dispersion of ant lion *Glenuroides japonicus* by Morisita and also his phenomenological theory [4] gave us just basically useful information. By generalizing Morisita's idea, we can derive the formula of a nonlinear diffusion which describes the density dependent dispersive motion in the nonuniform environment (Shigesada, Kawasaki and Teramoto [9]). The formula can also be inspected by comparing the results with the data of some other experiments (Shigesada [10]).

MODIFICATION OF MORISITA'S IDEA

The experiment of ant lions was done using a box, just one half of which is filled with fine sand and the other half is filled with coarse sand. The ant lions were put one by one on the border line of the two areas and the number of individuals settled in each type of sand was counted after their pit formation. It was found that when the population density is relatively low, the ant lions have a strong tendency to prefer fine sand to coarse sand for pit formation, but this tendency gradually falls with increasing density until an almost equal number of individuals settle their residences in both types of sand. This fact reflects the effect of repulsive interference among the individuals.

In order to explain this experimental result quantitatively, Morisita assumed that the probability of settlement of an individual in fine sand area A (or coarse sand area B) is inversely proportional to the degree of "unfavorableness" of that habitat, and the degree of unfavorableness is given by the sum of the quantity E_A (or E_B) named "environmental density," which is a measure of the intrinsic unfavorableness of the habitat itself, and the number of individuals n_A (or n_B) already settled in the fine sand (or coarse sand) area.

Then the probabilities that an individual settles the residence and forms a pit in fine sand area A or coarse sand area B when n_A and n_B individuals have already settled in area A and B respectively are given by

$$\begin{aligned} P_A &= (E_B + n_B)/(E_A + E_B + n_A + n_B) \\ P_B &= (E_A + n_A)/(E_A + E_B + n_A + n_B) \end{aligned} \tag{13.1}$$

and analyzing this Markov process, the average number of individuals found in the fine sand area, after n individuals are introduced into the box, can be obtained as

$$\frac{\overline{n_A}}{n} = \frac{E_B + (n - 1)/2}{E_A + E_B + n - 1} \tag{13.2}$$

It is seen that this ratio approaches a value 1/2, as n increases. The experimental results were quantitatively explained by this formula successfully.

In order to give an ethological interpretation to this phenomenon, here we assume that the movement of an individual in search of residence is under the influence of the following three forces: i) the dispersive force which is associated with the random movement of individuals, ii) the population pressure due to mutual interferences between individuals and iii) the attractive force toward the favorable places. The population pressure may be taken into account by considering that the dispersive force is strengthened by the repulsive interference with the increase of population density. Then we can assume the isotropic dispersive force, which is the sum of the forces i) and ii), to be given by a dispersion coefficient expressed as

$$\alpha + \beta n(x)$$

where n(x) is the population density at the position x. As for the measure of "favorableness," we can introduce a quantity -U(x), where the negative sign is attached so as to regard the function U(x) as an environmental potential function. Thus the mean velocity of the movement caused by the favorableness of habitat is assumed to be proportional to the force produced by the potential function U(x), that is $-\mathrm{grad}_x U(x)$.

Now let us consider again the experiment of ant lions. When a new individual is put in and is searching the residence in the box in which n_A and n_B individuals have already settled in A and B areas respectively, the transition probability rate at which the added individual transfer from B area to A area is considered to be given by $\alpha_B + \beta_B n_B + \kappa/2$, where κ is proportional to the potential difference $U_B - U_A$ which is positive in this case since the value of potential in the fine sand area A is lower than that of the coarse sand area B. Similarly, the transition probability rate from A to B is given by $\alpha_A + \beta_A n_A - \kappa/2$. Then it can be easily shown that the probability of settlement of a newly added ant lion in A area, P_A, becomes proportional to the transition probability rate from B to A and the same relation also holds for P_B, thus we have

$$\begin{aligned} P_A &= (\alpha_B + \beta_B n_B + \kappa/2)/(\alpha_A + \alpha_B + \beta_A n_A + \beta_B n_B) \\ P &= (\alpha_A + \beta_A n_A - \kappa/2)/(\alpha_A + \alpha_B + \beta_A n_A + \beta_B n_B) \end{aligned} \qquad (13.3)$$

This is nothing but a modified interpretation of Morisita's idea, so the expression (13.3) becomes equivalent to the Equation (13.1) by equating the environmental densities as

$$E_A = (\alpha_A - \kappa/2)/\beta, \qquad E_B = (\alpha_B + \kappa/2)/\beta \qquad (13.4)$$

where we assumed $\beta_A = \beta_B = \beta$. The average number of individuals found in A area after n individuals are introduced are given by, assuming $\alpha_A = \alpha_B = \alpha$,

$$\frac{\overline{n_A}}{n} = \frac{(\alpha + \kappa/2)/\beta + (n - 1)/2}{2\alpha/\beta + n - 1} \qquad (13.5)$$

As the next step in the derivation of a nonlinear diffusion equation, we shall consider the following experiment. Instead of introducing the individuals one by one, if we put all n individuals simultaneously into the box, they move around in the box under the influence of mutual interferences and the transfer probability rate from B to A is again given by $\alpha + \beta n_B + \kappa/2$ and similarly from A to B by $\alpha + \beta n_A - \kappa/2$, where the relation $n_A + n_B = n$ holds all through the process. Then, the temporal change of the number of individuals in each area can be represented by the equations

$$\frac{dn_A}{dt} = (\alpha + \beta n_B + \kappa/2)n_B - (\alpha + \beta n_A - \kappa/2)n_A$$

$$\frac{dn_B}{dt} = (\alpha + \beta n_A - \kappa/2)n_A - (\alpha + \beta n_B + \kappa/2)n_B \tag{13.6}$$

The solution can be again easily obtained analytically, and it is seen that the distribution always approaches a stationary distribution given by

$$\frac{\overline{n_A}}{n} = \frac{(\alpha + \kappa/2)/2\beta + (n - 1)/2}{\alpha/\beta + (n - 1)} \tag{13.7}$$

Here it will be seen that if we replace β in the equation (13.5) by 2β, it becomes equivalent to the expression (13.7). This is due to the fact that, in the latter case, each individual is under the influence of all other individuals the whole time.

SPACE CONTINUOUS MODEL

Here we can consider a similar experiment in which we use a linear array of many habitats which have different environmental conditions, for instance different particle sizes of sand for the ant lion. Then, in just the same way as Equations (13.6) for the case of two habitats, we can obtain a set of equations each of which describes the temporal change of the number of individuals n_i in the *ith* habitat [9].

In order to find the formula for the space continuous model, we can take the continuous limit of the equations for this discrete array model in a straightforward way. Then it can be shown that the population flow at the position x is given by

$$J = -\mathrm{grad}_x\{(\alpha + \beta n(x, t))n(x, t)\} - n(x, t)\mathrm{grad}_x U(x) \tag{13.8}$$

where $n(x, t)$ is the population density at the position x and time t. The first and second terms of Equation (13.8) correspond to the density dependent dispersion and the flow due to the environmental potential respectively. Using this flow function, we can express the equation for the change of density distribution by the continuity equation

$$\frac{\partial}{\partial t} n(x, t) = -\mathrm{div}\, J \tag{13.9}$$

Here we consider a one-dimensional box with boundaries at which $J = 0$. A stationary distribution $n^*(x)$ is given as a solution of the equation $J = 0$, that is

$$(\alpha + 2\beta n^*) \frac{dn^*}{dx} + \frac{dU}{dx} n^* = 0 \tag{13.10}$$

and by integrating this equation we have

$$2\beta\{n^*(x) - n(0)\} + \alpha \ln\{n^*(x)/n(0)\} = -\{U(x) - U(0)\} \tag{13.11}$$

where n(0) and U(0) are the value at the origin of the co-ordinate arbitrarily chosen in the habitat area.

Furthermore it can be proved that the stationary solution (13.11) is globally stable stationary distribution, by showing that the function

$$H = \int\{\alpha n \ln(n/n^*) - \alpha(n - n^*) + \beta(n - n^*)^2\}dx \geqq 0 \tag{13.12}$$

is the Lyapunov function of the equation (13.9). Actually the time derivative of the function H is given by

$$\frac{dH}{dt} = -\int(J^2/n)dx \leqq 0. \tag{13.13}$$

As an example, we can have the result given by Figure 13.1 which shows the stationary distributions when the environmental potential function is given by $U(x) = 1.5(x - 1)^2$. We can see again the characteristic feature of Morisita's experimental result that the distribution becomes flatter with increasing the total number of individuals N or the population pressure β.

INSPECTION BY OTHER EXPERIMENTS

Plausibility of the present formulation can be inspected by applying our nonlinear diffusion equation (13.9) to other experimental data of the dispersive motions of

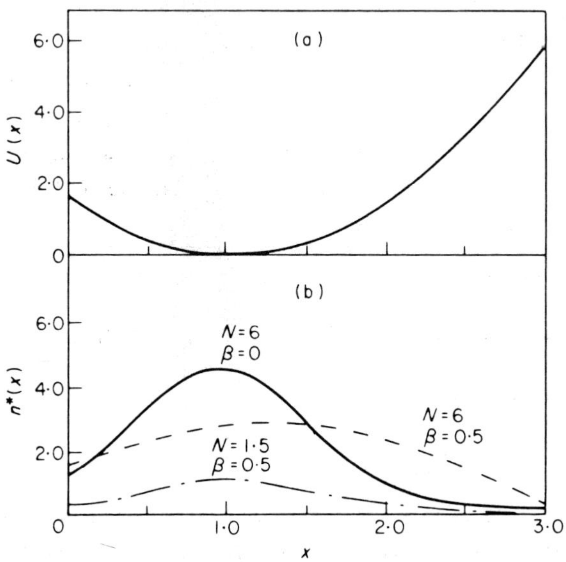

Fig. 13.1. (a) Environmental potential function $U(x) = 1.5(x - 1)^2$. (b) Stational population densities. N is the total number of individuals.

insects studied also by Morisita for ant lions [6], by Watanabe, Utida and Yosida for azuki bean weevils [13] and by Clark for grasshoppers [1].

In these experiments a number of individuals are placed in the center of experimental field with uniform environmental condition. The released individuals migrate randomly in the field until they eventually settle their residence. In this process, we can observe the spread of the distribution quantitatively as a function of time, by counting the number of individuals $F(r, t)$ found inside a circle of radius r centered at the releasing point, or by evaluating the variance of the distribution of the whole population $\sigma^2(t)$. In order to analyze the experimental data for these quantities, we can apply Equation (13.9) to the two-dimensional dispersive motions of the insects in the uniform field [10]. We assume that N individuals released at the origin at time $t = 0$ disperse by random movement and become sedentary somewhere at a constant rate μ. We denote the population densities of migratory and sedentary insects by $n(r, t)$ and $s(r, t)$ respectively as the function of the distance r from the origin and the time t. Then, in a two-dimensional uniform field, our nonlinear diffusion equation (13.9) can be expressed in terms of polar coordinates as

$$\frac{\partial}{\partial t} n(r, t) = \frac{1}{r}\frac{\partial}{\partial r}\left[r \frac{\partial}{\partial r}\{\alpha + \beta n(r, t)\}n(r, t)\right] - \mu n(r, t) \tag{13.14}$$

$$\frac{\partial}{\partial t} s(r, t) = \mu n(r, t) \tag{13.15}$$

For mathematical convenience, here we assumed that the dispersion coefficient depends only upon the density of migratory individuals $n(r, t)$. This assumption may be admitted only when the migratory individuals suffer smaller interference from the sedentary individuals as compared with that from migratory insects. Furthermore it is known from the experimental data that α is usually relatively small compared with $\beta n(r, t)$ except in peripheral regions, so, as a first approximation, we assume $\alpha = 0$ in the present analyses.

Then, using the transformation ([3], [10])

$$\tau = \frac{1}{\mu}(1 - \exp(-\mu t)), \qquad \xi(r, \tau) = n(r, t)\exp(\mu t)$$

Equation (13.14) can be reduced to

$$\frac{\partial}{\partial \tau} \xi(r, \tau) = \frac{1}{r}\frac{\partial}{\partial r}\left[r \frac{\partial}{\partial r} \beta\xi(r, \tau)^2\right] \tag{13.16}$$

which was already solved analytically by Pattle [7]. Thus it can be shown that the solutions of Equations (13.14) and (13.15) with the given initial conditions are given by

$$n(r,\ t) = \frac{2N\exp(-\mu t)}{\pi R(t)^2}\left[1 - \frac{r^2}{R(t)^2}\right] \quad \text{for } r < R(t)$$
$$= 0 \quad \text{for } r \geqq R(t) \tag{13.17}$$

$$s(r,\ t) = \frac{\mu R(t)^2}{8\beta}\left[1 - \frac{r^2}{R(t)^2} + 2\frac{r}{R(t)}\ln\frac{r}{R(t)}\right] \quad \text{for } r < R(t)$$
$$= 0 \quad \text{for } r \geqq R(t) \tag{13.18}$$

where

$$R(t) = \{32\beta N(1 - \exp(-\mu t))/\pi\mu\}^{1/4} \tag{13.19}$$

Equation (13.19) represents the dispersal front which separates the populated and unpopulated domains, and it increases monotonically with time and tends to a limiting value $R^* = \{32\beta N/\pi\mu\}^{1/4}$.

Now using these solutions we can evaluate the number of individuals $F(r,\ t)$ found inside a circle of radius r and also the variance $\sigma^2(t)$ of the distribution of the whole population by calculating

$$F(r,\ t) = \int_0^r \{n(r,\ t) + s(r,\ t)\}2\pi r\ dr \tag{13.20}$$

$$\sigma^2(t) = \frac{1}{N}\int_0^\infty r^2\{n(r,\ t) + s(r,\ t)\}2\pi r\ dr \tag{13.21}$$

and finally we have

$$F(r,\ t) = N\left[2\frac{r^2}{R(t)^2} - \frac{r^4}{R(t)^4} + 2(1 - \exp(\ \mu t)) \times \left\{\frac{r^2}{R(t)^2} - \frac{r^4}{R(t)^4} + 2\frac{r^4}{R(t)^4}\ln\frac{r}{R(t)}\right\}\right] \quad \text{for } r < R(t)$$
$$= 0 \quad \text{for } r \geqq R(t) \tag{13.22}$$

$$\sigma^2(t) = \frac{8}{9}\left[\frac{2\beta N}{\pi\mu}\{1 - \exp(-\mu t)\}\right]^{1/2}\{1 + \frac{1}{2}\exp(-\mu t)\} \tag{13.23}$$

DISCUSSION

It seems to be worthwhile to compare the present results with those for the case of density independent dispersion with a constant diffusion coefficient α and $\beta = 0$. In this case one can obtain

$$n(r,\ t) = \frac{N}{4\pi\alpha t}\exp(-r^2/4\alpha t - \mu t) \tag{13.24}$$

$$s(r,\ t) = \int_0^t n(r,\ t')dt \underset{t\to\infty}{\rightarrow} \frac{\mu N}{2\pi\alpha}K_0((\mu/\alpha)^{1/2}r) \tag{13.25}$$

$$F(r, t) = N\{1 - \exp(-r^2/4\alpha t - \mu t) - \mu \int_0^t \exp(-r^2/4\alpha t' - \mu t')dt'\}$$

$$\underset{t\to\infty}{\to} N\{1 - (\mu/\alpha)^{1/2} r K_1((\mu/\alpha)^{1/2} r)\} \tag{13.26}$$

$$\sigma^2(t) = \frac{4\alpha}{\mu}\{1 - \exp(-\mu t)\} \tag{13.27}$$

where K_0 and K_1 are the modified Bessel functions of zero and 1st kinds respectively.

Contrarily to the results given by Equations (13.17), (13.18) and (13.22) which have the distribution front R(t), the characteristic feature of these distribution functions (13.24), (13.25) and (13.26) is that they have tails extending to infinity. For instance, the density distribution of sedentary insects s(r, t) at time $t \to \infty$ is shown in Figure 13.3 which gives a curve tending to infinity as $r \to 0$. On the contrary, the present result given by Equation (13.18) has a finite population density at r = 0 as shown in Figure 13.2.

Figure 13.4 shows the distribution $F(r, \infty)$ in the final state when all the insects are sedentarily settled. Experimental data were given also by Morisita and the theoretical curves are plotted by the equation

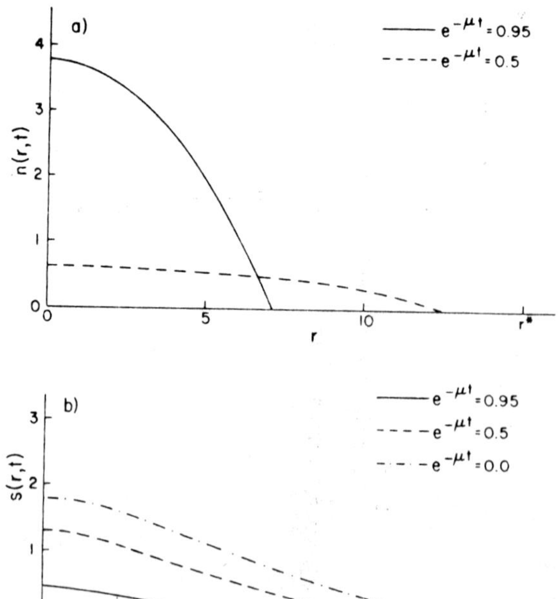

Fig. 13.2. Density distribution of insects population. (a) dispersing insects. (b) sedentary insects. Parameters chosen are N = 100; $R(t) = 15(1 - \exp(-\mu t))$.

$$F(r, \infty) = N \left\{ 4 \frac{r^2}{R^{*4}} + 3 \frac{r^4}{R^{*4}} + 4 \frac{r^4}{R^{*4}} \ln \frac{r}{R^*} \right\} \qquad r < R^*$$
$$= N \qquad\qquad r \geqq R^* \tag{13.28}$$

which is obtained from Equation (13.22) by taking the limit $t \to \infty$.

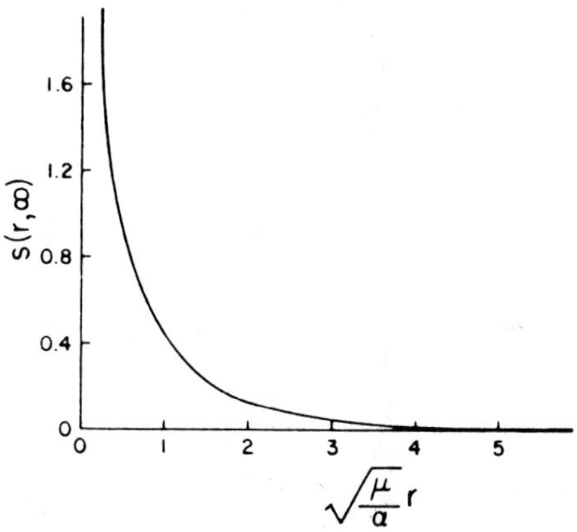

Fig. 13.3. Ultimate density distribution of sedentary animals in the absence of population pressure. Parameters are chosen so that $\mu N/2\pi\alpha = 1$.

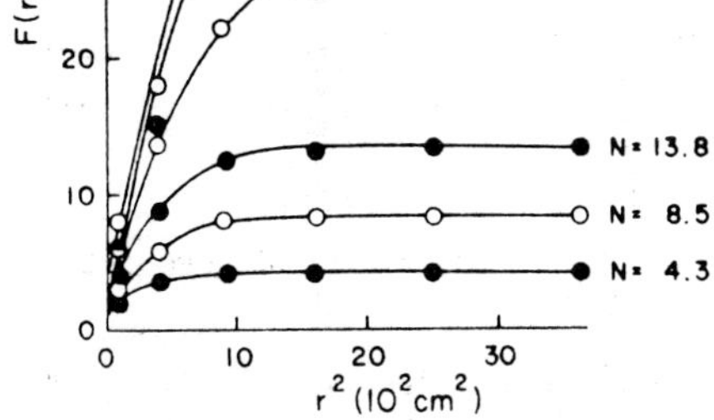

Fig. 13.4. Number of ant lions settled inside a circle of radius r. Open and solid circles are experimental data for various numbers of ant lions N [6]. Solid lines are theoretical curves calculated from Eq. (13.28) under the condition $R^* = 21N^{1/4}$.

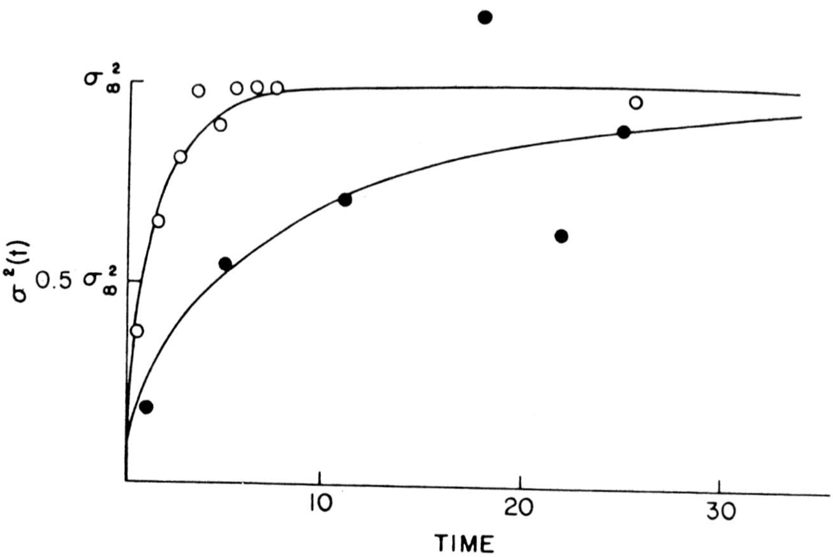

Fig. 13.5. Temporal change in variance of animal distributions. Open circles: experimental data of azuki bean weevils obtained by Watanabe et al. The time scale is in hours. Solid circles: data for grasshoppers obtained by Clark [1]. The time scale is in days. Solid lines are theoretical curves calculated from Eq. (13.23) in which the parameter μ is chosen as 0.2/hour for the upper curve and 0.03/day for the lower curve.

As for the transient stage of dispersal movement, it is seen that the variance of the distribution $\sigma^2(t)$ given by (13.23) shows a fairly good fit with the data given by Watanabe et al. for azuki been weevils and by Clark for grasshoppers.

Unlike the passive dispersive motions due to the physical external forces such as the dispersion of the spores and phytoplanktons, the characteristic features of the active dispersive movement of animals have the variety of motional patterns depending upon the species characteristics. In the present discussions, we study only the simplest case in which the isotropic dispersive movements are caused by the population pressure. Therefore, the applicability of our formulation is inevitably restricted to the dispersion of animals having rather simple behavioral patterns.

REFERENCES:

1. Clark, D. P., An analysis of dispersal and movement in Phaulacridium vittatum, Aust. J. Zool. 10: 382 (1962).

2. Einstein, A., Uber die von der molekularkinetischen Theorie der Warme geforderte Bewegung von in ruhenden Flussigkeiten Suspendierten Teilchen, Ann. der Physik 17: 549 (1905).

3. Gurtin, M. E. and MacCamy, R. C., On the diffusion of biological populations, Math. Biosciences 33: 35 (1977).

4. Morisita, M., Habitat preference and evaluation of environment of an animal: experimental studies on the population density of an ant-lion, Glenuroides japonicus M'L (1) Physiol. and Ecol. 5: 1 (1952).

5. Morisita, M., On the relationship between dispersal of ant-lion and its population density, Japanese J. Zool. 63: 22 (1954).

6. Morisita, M., Dispersion and population pressure: experimental studies on the population density of an ant-lion, Glenuroides japonicus M'L (2), Japanese J. Ecol. 4: 71 (1954).

7. Pattle, R. E., Diffusion from an instantaneous point source with a concentration-dependent coefficient, Quart. J. Mech. Appl. Math. 12: 407 (1959).

8. Shigesada, N. and Teramoto, E., A consideration on the theory of environmental density, Japanese J. Ecol. 28: 1 (1978) (Japanese with English summary).

9. Shigesada, N., Kawasaki, K. and Teramoto, E., Spatial segregation of interacting species, J. Theor. Biol. 79: 83 (1979).

10. Shigesada, N., Spatial distribution of dispersing animals, J. Math. Biol. 9: 85 (1980).

11. Smoluchowski, M. v., Zur kinetischen Theorie der Brownschen Molekularbewegung und der Suspensionen, Ann. der Physik 21: 756 (1906).

12. The Svedberg and K. Inouye, Eine neue Methode zur Prufung der Gultigkeit des Boyle-Gay-Lussacschen Gesetzes fur Kolloide Losungen, Zeitschr. f. Physik. Chemie 73: 547 (1910); 77: 145 (1911).

THE NEUROSPORA MYCELIUM AS A TWO-DIMENSIONAL CONTINUUM OF COUPLED CIRCADIAN CLOCKS

A. T. Winfree and G. M. Twaddle
Purdue University, Department of Biological Sciences,
West Lafayette, Indiana

ABSTRACT

While advancing across the surface of a nutrient agar, the frontier of a mycelium of the fungus NEUROSPORA CRASSA alternates at circadian intervals between purely vegetative growth and differentiation of asexual spores. Though expressed only at the frontier, this circadian oscillation can be shown to continue in older hyphae far behind the frontier. Using local exposures to blue light to differentially phase shift adjacent areas of oscillating mycelium, we investigated the presumed coupling and mutual synchronization of nearby hyphae. We found no such effects. Artificially created phase discontinuities (1/2 cycle within 1 cm.) and shallower phase gradients (1 to 2 hours per cm.) persisted without detectible alteration for at least 5 cycles.

We encountered some technical difficulties. The bd-csp-1a strain was found to be a "stopper" beyond a week's growth. Also phase gradients exceeding 2 hours per cm. spontaneously fragmented into shallower gradients separated by abrupt discontinuities unpredictably located. We believe this to be a consequence of large irregular and temporary fluctuations of growth rate combined with competition among hyphae on the frontier.

INTRODUCTION

Dark-grown cultures of the bd strain of NEUROSPORA CRASSA exhibit a circadian rhythm in the periodic formation of conidia (asexual spores) [20]. As a mycelium grows, at intervals of 21 hours conidia form in bands along an advancing frontier of vegetative hyphae. After a few days of growth a record of the circadian rhythm and of past frontier positions is preserved in the pattern of alternating conidial and vegetative bands.

This rhythm is typically circadian in that its period is relatively independent of constant temperature, it is phase-shifted by changes of temperature and can be entrained by temperature cycles or suppressed by constant cold [11], it is phase-shifted by brief light pulses and can be entrained by light/dark cycles or suppressed by constant light ([18], [20]).

Investigations into the molecular mechanisms underlying this rhythm have revealed circadian fluctuations in a number of biochemical parameters ([2], [3], [4], [13], [15]). In other studies specific drug-pulses have been used to uncover cellular processes essential to the rhythm ([8], [12], [16]). Genetic studies have isolated and mapped specific period mutants ([7], [8], [9], [16]). Yet, like the circadian rhythms of all other organisms, the molecular mechanism of the conidiation rhythm in NEUROSPORA remains unknown.

NEUROSPORA is a unique model for the study of circadian clocks in that it permits observation of the spatial patterning (if any) of circadian timing. The circadian rhythm is initially expressed only along the one-dimensional growing frontier, while back in the older two-dimensional expanse of mycelium there are no conspicuous morphological changes. Yet in a recent study Dharmananda and Feldman [5] discovered that circadian time-keeping persists within the older mycelium. They demonstrated that patches of hyphae, cut from anywhere in a mycelium and transferred to fresh nutrient agar, would exhibit circadian conidiation in phase with the rhythm observed on the frontier of the original mycelium. Closer work revealed a slight gradient of phase from frontier to interior of an old mycelium, suggesting that frontier hyphae oscillate slightly faster than those left behind, and that regions within a mycelium need not maintain strict synchrony. Thus NEUROSPORA appears to be a two-dimensional continuum of locally autonomous circadian clocks, which can now be manipulated by local stimuli.

Our purpose was to investigate the stability of patterns of clock phase within a mycelium. Does the lack of complete synchrony indicate complete autonomy of hyphal clocks, or is there some communication of phase information, at least over short distances after a few days of contact? Normally communication is not observable because all points on the frontier are at the same circadian phase. But by local phase-shifting we can create phase gradients and phase discontinuities along the frontier. Conidiation bands thereafter are not parallel to the frontier: their shapes reveal the timing gradient and changes in their shapes reveal changes of timing between adjacent patches of frontier. If NEUROSPORA CRASSA is a continuum of coupled circadian oscillators, when viewed on the gross scale of days and centimeters, then we should be able to measure the rate of transfer of phase information throughout the mycelium. If that rate is large then there may be interesting consequences such as propagating waves of circadian resetting (if clock kinetics is excitable, like a relaxation oscillator), or mutual synchronization in patches analogous to magnetic domains, or phase-diffusion at a rate traceable to translocation of a mobile component of the cytoplasm. And if the transfer rate of phase information should prove to be small, then many subcellular organelles may be excluded as carriers. A mycelium without significant internal coupling would also allow convenient measurements on phase- and amplitude-resetting in two-dimensional gradients that combine all phases and all stimulus magnitudes within one mycelium, as an alternative to using a separate inoculum and race-tube for each combination.

MATERIALS AND METHODS

The bd-csp-1a strain of NEUROSPORA CRASSA used in this study was provided by Dr. Jerry Feldman. The band (bd) allele enables expression of circadian rhythmicity even at the fairly high CO^2 tensions typical of petri dish culture ([19], [20], [24]). The conidial separation allele (csp) prevents detachment of conidia from hyphae, so transferred strips of mycelium do not reinoculate their environs with spores [21]. Our cultures were grown on 2% agar (to facilitate manipulations) containing Vogel's salts [27], 1.2% sodium acetate, and 0.02% casamino acids (Feldman's recipe). Stock cultures were kept on slants at 4 degrees C and transferred every month to new slants.

For experiments spores were germinated on nutrient agar in 15 cm. petri dishes under blue "moonlight" (see below). Mycelia were allowed to grow for several days, then cut into thin strips to serve as inocula for 35 cm. by 18 cm. by 4 cm. plexiglass trays with non-reflecting plate glass tops. These contained a 3 mm. depth of the same nutrient agar, this depth being found to allow less than a day's growth at each spot before depletion or staling so the conidial banding remained distinct. Sufficient space was left for adequate gas exchange between the tray tops and their plate glass lids. (We measured complete respiration of 1/3 $cm.^3$ of air per $cm.^2$ of growing mycelium per hour.) Temperature was maintained at 25 degrees C throughout the experiment by keeping the trays (in an incubator except during light exposures) on water-cooled plates fed by a Lauda temperature-regulating circulator. The room was kept at a slightly higher temperature to minimize condensation on the glass lids. It should be noted that temperature changes as small as 5 degrees for 6 hours have a severe impact on circadian phasing in NEUROSPORA [11]. We monitored the temperature with min/max recording thermometers, reset daily.

Rhythmicity in the mycelium was suppressed by 4-5 days' growth under dim blue "moonlight" projected 350 cm. from a 1 mm. incandescent filament filtered through copper sulfate solution and a Wratten #47 celluloid. The long distance and small filament gave us sharp shadows and uniformity of exposure over the entire growth area. Intensity at the agar surface was 100 $ergs/cm.^2/sec.$ (0.1 $watt/m^2$). This sufficed to fully suppress rhythmicity within several days while not scattering functionally detectable amounts of light into adjacent shadowed areas.

The darkroom was otherwise illuminated by a red safelight which, even in prolonged exposures and/or at much higher intensity had no discernable effect on the conidiation rhythm. Both lights were the same as used for the past 12 years in this laboratory for phase resetting experiments with DROSOPHILA MELANOGASTER and PSEUDOOBSCURA [26]; to within a factor of two the fungus' photosensitivity for resetting and suppression seem the same as the flies'.

The circadian oscillation begins upon transferring to constant darkness (or red safelight) a mycelium whose rhythmicity has been suppressed by prior prolonged exposure to constant blue moonlight. By shadowing only half of the mycelium for 10.5 hours before additionally shadowing the other half, we were able to construct 1/2-cycle phase discontinuities. Phase gradients were formed similarly by moving an eclipsing shadow at constant speed along the advancing frontier. As light-to-dark transitions occur at successive points along the frontier, clock oscillations begin at correspondingly staggerred times (Figure 14.1). Gradients spanning 1/2 cycle to 3 full cycles on the 18 cm. frontier were formed by pulling the shadow at various speeds (the slower the eclipse, the greater the change of clock phase per cm., thus the steeper the gradient). The eclipsing card was later changed to a superposition of orange and red celluloids which, while blocking out the blue light, still allowed us to see and photograph the gradient taking shape under the shadow.

RESULTS

During 6-8 days of growth there was no detectable transfer of phase information in any of the spatial patterns examined. Artificially induced phase discontinuities (at least 1/2 cycle change per cm.) as well as those arising spontaneously (see below) and shallow phase gradients (as little as 1-2 hours per cm.) all persisted for at least 5 cycles without detectable systematic change in the shapes of newly-forming conidiation bands.

The steeper phase gradients formed spontaneous phase discontinuities at unpredictable locations. These discontinuities typically corresponded to an indentation on the frontier (Figure 14.2) and in at least one case arose immediately behind a contaminant plaque (Figure 14.3).

Mycelia in the large trays grew in a locally fanlike fashion with multiple hyphal branching and competition for space among adjacent regions. We believe that segments of the frontier may often have fluctuated to slower growth while adjacent segments temporarily grew faster, resulting in extinction of the less-competitive segment. This impression is reinforced by careful perusal of the fine texture and orientation of hyphae in photographs such as Figures 14.2 and 14.3. Some such fluctuations must be expected in general. Our impression is that they are of

Fig. 14.1. A comparison of the patterns of circadian clock phase in a) normal synchronous mycelia, b) mycelia with a phase discontinuity, and c) mycelia with a phase gradient. The arrow indicates the direction of growth. The circadian cycle is arbitrarily divided into four parts: conidiation is induced in hyphae forming at the frontier when the clock strikes 4. Each digit tells the local clock phase when that spot was on the frontier; the numbers accordingly increase regularly along every hyphal growth path, but the phasing of adjacent paths varies in cases a), b), and c).

surprising magnitude and duration, and that the extinguished areas are consequently large. This is used below to rationalize the random spontaneous emergence of phase discontinuities preferentially in steeper phase gradients.

In this strain, cultured under all the diverse conditions we tried, still larger fluctuations of growth rate severely limited our observations. Anywhere from 5 to 10 days after inoculation the frontier would (apparently "spontaneously") stop growing in isolated segments. These eventually spread to engulf the entire frontier,

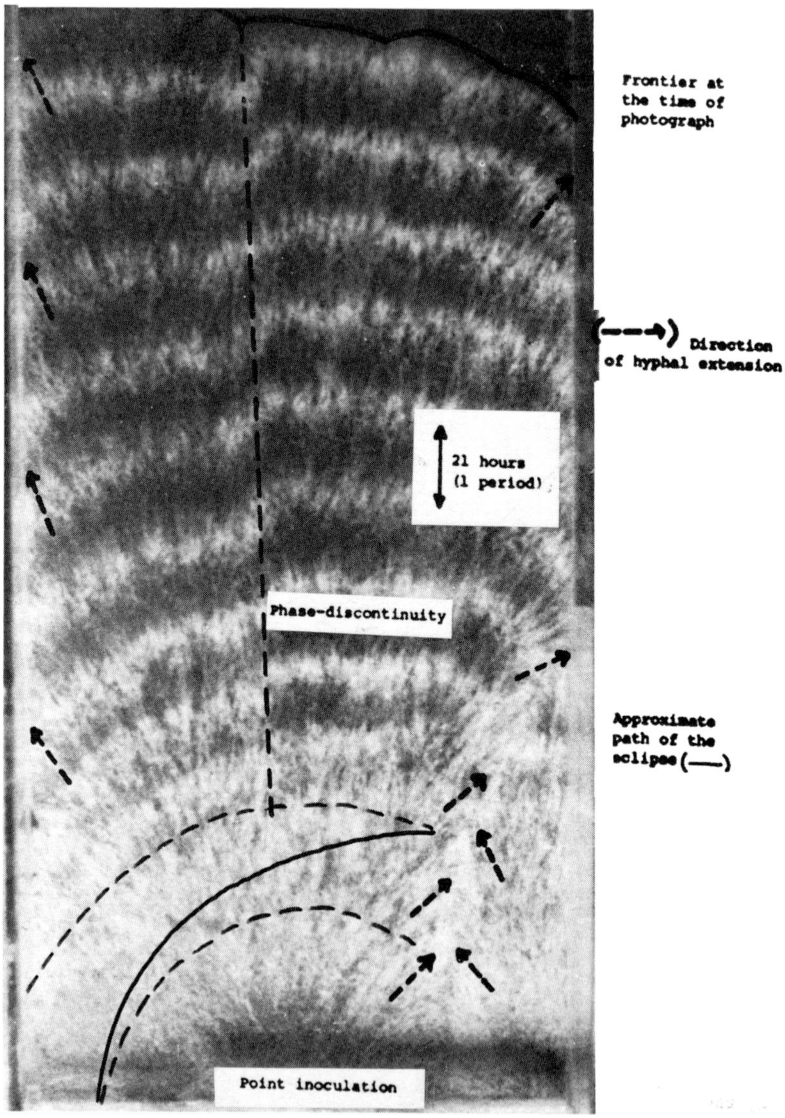

Fig. 14.2. A mycelium grown from a point inoculum. The direction of hyphal extension is shown by arrows. A steep phase gradient was formed by drawing an eclipse from left to right. The broken curves show where the frontier was at the beginning and end of the eclipse; the solid curve traces the approximate path of the eclipse along the moving frontier. Note that conidiation bands are initially parallel to the latter (not to the frontier). A phase discontinuity soon arises along the straight dashed growth path.

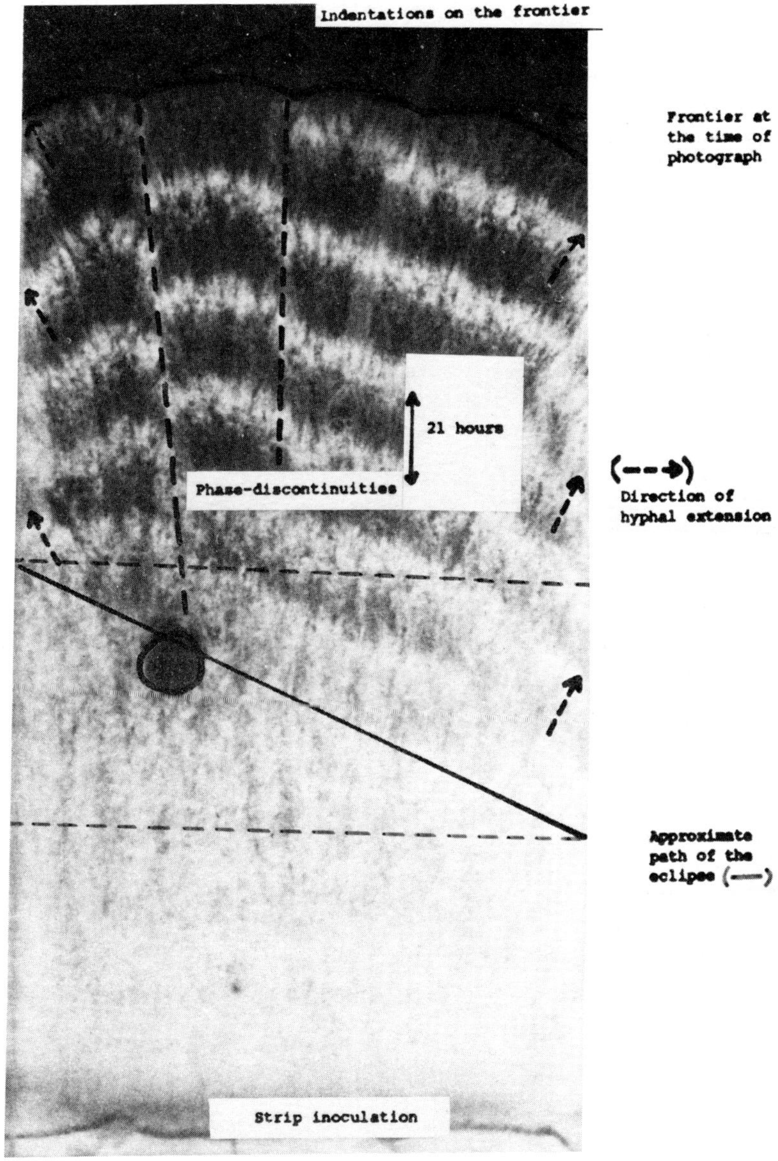

Fig. 14.3. Similar to Figure 14.2 except that the inoculum was a strip of mycelium rather than a dot of spores, and the eclipse went from right to left. One phase discontinuity arises from an area of contamination and another arises spontaneously during the third cycle after the eclipse.

even though the agar immediately ahead remained perfectly capable of supporting normal growth. After extensive experiments which finally exhausted every reasonable interpretation we could think of, we were fortunate to encounter T. H. Pittenger, who told us about "stopper" strains [1]. For reasons that remain obscure, such strains repeatedly cease growth for days, weeks, or even months at a time and then resume ostensibly normal growth. Apparently bd-csp-1a is one such. Dr. S. Brody's bd-csp-2A proved not to have this quirk, but we were unable to optimize culture conditions for 2A as well as for 1a and did not obtain sufficiently distinct banding for our shape measurements.

DISCUSSION

An ascomycete such as NEUROSPORA is not strictly a cellular organism. Hyphal segments are divided by porous septae that allow free translocation of cytoplasm and organelles. Translocation rates may vary enormously from strain to strain and with age and position in the mycelium ([14], [22], [23]). The 5-9 cycles through which phase discontinuities and phase gradients were observed to persist would seem more than enough time for translocation of small molecules throughout 1 $cm.^2$ of mycelium. Biochemical exchange on that scale of time and space could be driven even by diffusion alone, supposing the pertinent clock quantity comparable in size to glucose (with diffusion coefficient about 10^{-5} $cm.^2$ per second or 1 $cm.^2$ per day in water). As we observed no alteration of phase patterns down to a resolution of about 1 cm. in several days, we think there was virtually no transfer of phase information. Perhaps the clock mechanism is compartmentalized in organelles too big to pass septal pores or is in some way bound to immovable membranes within the hyphae.

The instability of steep phase gradients initially impressed us as an indication of communication along the frontier. However they arise apparently at any random place during any cycle after the initially-smooth gradient is established. We now believe the instability is a consequence of competition among hyphae growing on the frontier. If the agar ahead is colonized, not equally by all hyphae, but selectively by segments that are growing faster for a while, then intervening segments may be out-competed and excluded from representation along the later, further advanced frontier. The corresponding range of clock phases would be excluded with them (Figure 14.4). In steeper gradients that range would be larger (in hours per lateral cm., thus in forward cm. per lateral cm.). In synchronous mycelia there would be no observable consequence at all. Thus circadian phase gradients may provide a first measure of the extent of competitive fluctuations in colonial fungi.

This interpretation suggests an explanation for the characteristic indentation of the frontier at the discontinuity (Figures 14.2 and 14.3). If a segment of frontier is being squeezed out by the angling-inward of hyphae from both sides, then

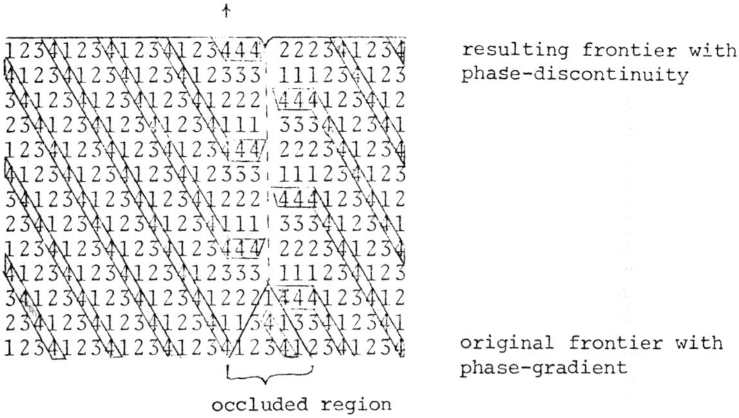

Fig. 14.4. An elaboration of Figure 14.1c to indicate how lateral competition along the frontier can occlude a segment of a smooth phase gradient, resulting in a phase discontinuity. Note that the discontinuity is necessarily straddled by zones of synchrony to fill the original space. In these regions, unlike the adjacent gradient and discontinuity, conidiation should be parallel to the frontier, but the difference is hard to see in gradients as shallow as ours.

the frontiers of growth will be slightly tilted on both sides as they come together, leaving a notch where they collide. It is also easy to see how an obstacle such as a contaminant colony, by impeding the growth of hyphae that directly encounter it, might allow neighbors on both sides to occlude the retarded segment and thus delete its range of phases from representation on the frontier. Figure 14.3 seems to show such an event.

Since the hyphae grow at about 1 mm./hour independent of clock phase, there is a minimum time required for complete encroachment from both sides into a non-growing segment big enough to constitute a noticeable discontinuity. This time is related to the steepness of the phase gradient by the expression: $t = G\Delta\phi/2g$ where $\Delta\phi$ is the change of phase spanned by the missing segment (the discontinuity), g is the growth rate (0.1 cm./hour), and G is the speed of the eclipse along the frontier (cm./hour). Earlier appearance of a phase discontinuity of this size would discredit our interpretation (supposing that contaminants and other irregularities have been eliminated). For a half-cycle discontinuity in a 24 mm./day gradient, t is in the order of 10 hours; no discontinuities have yet been noticed so early.

Within their limits, these experiments indicate substantial local autonomy in the cycling of these circadian clocks. They show that spatial resolution in the order of millimeters will be required for detection of communication within a few days. And they show that careful measurements of translocation rates will be necessary and worthwhile for discovering what parts of the hyphae support circadian mechanisms.

A recent genetic study [6] associates mitochondria with a period-shortening mutation; it would accordingly be good to find out the translocation mobility of mitochondria as a function of distance behind the frontier.

Another little observation that may be of interest in this connection concerns dark-reared mycelia. They usually exhibit normal banding in concentric rings about 21 hours apart, but each mycelium is randomly phased, independent of others simultaneously germinated in the same incubator. If hyphae were strictly autonomous in initiating their circadian rhythms, one might expect also to see SPIRAL banding, as does occur in NECTRIA CINNABARINA [25]. Morphogenesis would be constrained to ring patterns only if for some reason the initial phasing of local oscillations were so constrained that a quick tour of inspection around the frontier would lead one forward in phase exactly as much as backward, i.e., only if there could never be a net advance or delay of a complete cycle around the frontier [26]. This would be the case, for example, if there were strong enough metabolic coupling within the nascent mycelium so that the clock necessarily starts synchronously throughout its initially small area. However synchrony could also result, without communication, from fluctuations of growth rate in which a small portion of the young mycelium would become the progenitor for the whole colony. By inoculating agar with a sufficiently large thin ring of spores (instead of a point or small disk) it should be possible to observe the extent of spatial correlation in startup phase and to obtain spiral morphogenesis in NEUROSPORA as in other ascomycetes.

The absence of demonstrable coupling may offer a technical convenience for experimental studies of clock biochemistry, at least in strains whose forward and lateral growth rates are more stable than they seem to be in bd-csp-1a. We have successfully created smooth phase gradients spanning more than 1 full cycle in a baking-dish culture. If the gradient were stable, with regions 1 cm. apart essentially independent in their rhythmic activities, then exposure to a single perturbation (e.g., a flash of light, a temperature pulse, a brief transfer onto acetaldehyde-agar) would simultaneously measure phase resetting at each phase in the cycle. The resetting curve would be drawn directly on the mycelium as the next band of conidia. Given the fact of type 0 resetting (Feldman and Dharmananda personal communication, and repeats in our laboratory), a special trick recommends itself. The first step is to use type 0 resetting to create a circular phase gradient running one full cycle around the perimeter of a disk-shaped mycelium [26]. This would initiate spiral conidiation outside the original concentric rings of conidiation. One could then quantitate the impact of any perturbation by measuring the whole phase resetting curve in one shot on a single mycelium as noted above (but now in circular rather than rectilinear layout). The magnitude of that stimulus at which resetting abruptly switches from "weak" to "strong" would be signalled unmistakeably by a switch back to concentric ring morphogenesis.

Another unique feature of a closed circular phase gradient is that its interior necessarily harbors a singular point at which the clock is in a delicate state of indeterminate phase [26]. In all other experimental arrangements that we know of, the singularity proves to be a slippery, elusive state, possibly quite unstable. But in a two-dimensional continuum of connected clocks it can be trapped for as long a time as the clocks around the border continue to cycle normally, given initial phasing as described above. This time will be longer the bigger the mycelium and the weaker the phase communication within it. Given a trapped singularity one might then classify each type of perturbing stimulus (e.g., various metabolites or inhibitors) into one of several discrete groups by observing the physical direction (relative to the momentary location of phase zero on the border) in which tiny doses displace that singularity on the mycelium. In theory this direction should be independent of dose, very small doses should suffice, and the timing of the dose should not matter. Presumably the groups identified would correspond to the points in clock metabolism at which the various kinds of perturbation intervene. The ways in which they classify might be biochemically interesting.

The use of spatial patterning to enquire into circadian physiology is only beginning. This particular attempt was hampered by difficulties of technique specific to culture on a large scale and by our rediscovery of the "stopper" trait which apparently needs to be removed from the genetic background of bd-csp-1a. Further studies may also have to contend with hyphal competition on the frontier (resulting in occlusion of random segments) unless elimination of "stopper" drastically reduces the magnitude and duration of putative local fluctuations in growth rate.

ACKNOWLEDGMENTS

This effort was funded in part by the National Science Foundation through grant CHE 77-24649. Harold Aberman's diligence and imaginative attention to detail made much possible that looked hopeless.

REFERENCES:

1. Bertrand, H., McDougall, K. J. and Pittenger, T. H., Somatic variation during uninterrupted growth of NEUROSPORA CRASSA in continuous growth tubes, J. Gen. Microbiol. 50: 37-50 (1968).

2. Brody, S. and Harris, S., Circadian rhythms in NEUROSPORA: spatial differences in pyridine nucleotide levels, Science 180: 498-500 (1973).

3. Brody, S. and Martins, S. A., Circadian rhythms in NEUROSPORA CRASSA: effects of unsaturated fatty acids, J. Bacteriol. 137: 912-915 (1979).

4. Delmer, D. and Brody, S., Circadian rhythms in NEUROSPORA CRASSA: Oscillation in the level of an adenine neucleotide, J. Bacteriol. 121: 548-553 (1975).

5. Dharmananda, S. and Feldman, J. F., Spatial distribution of circadian rhythm in aging cultures of NEUROSPORA CRASSA, Plant Physiol. 63: 1049-1054 (1979).

6. Dieckman, C. and Brody, S., Circadian rhythms in NEUROSPORA CRASSA: Oligomycin resistance mutation affecting periodicity, Science 207: 896-898 (1980).

7. Feldman, J. F. and Hoyle, M. N., Isolation of circadian clock mutants of NEUROSPORA CRASSA, Genetics 75: 605-613 (1973).

8. Feldman, J. F., Circadian periodicity in NEUROSPORA: alteration by inhibitors of cyclic AMP phosphodiesterase, Science 190: 787-790 (1975).

9. Feldman, J. F. and Atkinson, C. A., Genetic and physiological characteristics of a slow-growing circadian clock mutant of NEUROSPORA CRASSA, Genetics 88: 255-265 (1978).

10. Feldman, J. F. and Hoyle, M. N., Complemention analysis of linked circadian clock mutants of NEUROSPORA CRASSA, Genetics 82: 9-17 (1976).

11. Francis, C. D. and Sargent, M. L., Effects of temperature perturbations on circadian conidiation in NEUROSPORA, Plant Physiol. 64: 1000-1004 (1979).

12. Frelinger, J. G., Motulsky, H. and Woodward, D. O., Effects of chloramphenicol on the circadian rhythm of NEUROSPORA CRASSA, Plant Physiol. 58: 592-594 (1976).

13. Hochberg, M. L. and Sargent, M. L., Rhythms of enzyme activity associated with circadian conidiation in NEUROSPORA CRASSA, J. Bacteriol. 120: 1164-1175 (1974).

14. Jennings, D. H., Thornton, J. D., Galpin, M. F. J. and Coggins, C. R., Translocation in fungi, Symp. Soc. Exp. Biol. 28: 139-156 (1974).

15. Martens, C. L. and Sargent, M. L., Circadian rhythm of nucleic acid metabolism in NEUROSPORA CRASSA, J. Bacteriol. 117: 1210-1215 (1974).

16. Sargent, M. L., The effects of four antibiotics on growth and periodicity of a rhythmic strain of NEUROSPORA, Neurospora Newsletter 15: 17 (1969).

17. Sargent, M. L. and Briggs, W. R., The effects of light on a circadian rhythm of conidiation in NEUROSPORA, Plant Physiol. 42: 1504-1510 (1967).

18. Sargent, M. L., Briggs, W. R. and Woodward, D. O., Circadian nature of a rhythm expressed by an intertaseless strain of NEUROSPORA CRASSA, Plant Physiol. 41: 1343-1349 (1966).

19. Sargent, M. L. and Kaltenborn, S. H., Effects of medium composition and carbon dioxide on circadian conidiation in NEUROSPORA, Plant Physiol. 50: 171-175 (1972).

20. Sargent, M. L. and Woodward, D. O., Genetic determinants of circadian rhythmicity in NEUROSPORA, J. Bacteriol. 97: 861-866 (1969).

21. Selintrennikov, C. P., Nelson, R. E. and Siegel, R. W., Phase-specific genes for macroconidiation in NEUROSPORA CRASSA, Genetics 78: 679-690 (1974).

22. Trinci, A. P. J., Growth of wild type and spreading colonial mutants of NEUROSPORA CRASSA in batch culture and on agar medium, Arch. Mikrobiol. 91: 113-126 (1973).

23. Trinci, A. P. J. and Collinge, A. J., Structure and plugging of septa of wild type and spreading colonial mutants of NEUROSPORA CRASSA, Arch. Mikrobiol. 91: 355-364 (1973).

24. West, D. J., Effects of pH and biotin on a circadian rhythm of conidiation in NEUROSPORA CRASSA, J. Bacteriol. 123: 387-389 (1975).

25. Winfree, A. T., Polymorphic pattern formation in the fungus NECTRIA, J. Theor. Biol. 38: 363-382 (1973).

26. Winfree, A. T., The Geometry of Biological Time, Springer Verlag, New York, 1980.

27. Vogel, H. J., Distribution of lysine pathways among fungi: evolutionary implications, Amer. Naturalist 98: 430-556 (1964).